THE PLASMA MEMBRANE

TRANSPORT IN THE LIFE SCIENCES

THE PLASMA MEMBRANE

S. K. Malhotra
University of Alberta

A Wiley-Interscience Publication

JOHN WILEY & SONS

New York · **Chichester** · **Brisbane** · **Toronto** · **Singapore**

Library of Congress Cataloging in Publication Data:

Malhotra, S. K. (Sudarshan Kumar), 1933–
    The plasma membrane.

    (Transport in the life sciences, ISSN 0271-6208; v. 6)
    "A Wiley-Interscience publication."
    Bibliography: p.
    Included index.
    1. Plasma membranes. I. Title. II. Series.
[DNLM: 1. Cell membrane—Physiology. 2. Cell membrane—
Ultrastructure. W1 TR235V v. 6 / QU 601 M249p]

QH601.M34   1983   574.87'5   83-5822
ISBN 0-471-09325-4

To My Parents and My Teachers

SERIES PREFACE

Membrane transport is rapidly becoming one of the best-worked fields of modern cell biology. The Transport in the Life Sciences series deals with this broad subject in monograph form. Each monograph seeks to trace the origin and development of ideas in the subject in such a way as to show its true relation to membrane function. It also seeks to present an up-to-date and readable outline of the main problems in the subject and to guide thought on to new lines of investigation.

The task of writing a monograph is not a light one. My special gratitude is to the various authors for expounding their subjects with scholarly care and force. For the preparation of the indexes I thank Dr. Barbara Littlewood.

E. EDWARD BITTAR

Madison, Wisconsin
June 1980

PREFACE

This monograph is an attempt at an elaboration of the current trends in the study of the structure and function of the plasma membrane. Such a discourse should be of interest to senior undergraduate and graduate students and postdoctoral fellows in biological and basic medical sciences. Teachers of cell biology may find it useful as supplementary reading material. Freshmen in general biology would find it worth their while going through it after their introductory courses in cell biology and biochemistry. Rapid advances are likely to take place in the field of the biology of membranes, particularly in the understanding of the functional organization of specific membrane-related phenomena. Nonetheless, this monograph will be useful as a general guide to the background literature for some years to come.

No attempt has been made to provide information on relevant tools and techniques or historical background and reviews of the vast and rapidly expanding literature available on various aspects of the plasma membrane or biological membranes in general. An attempt has been made to cover such topics on the structure and function of the plasma membrane as are likely to be common to a vast majority of the cells. Even though highly specific plasma membrane-associated structures such as acetylcholine receptors, acetylcholinesterase, and bacteriorhodopsin have been selected, the major features that emerge from the discussion of these topics are likely to be applicable to a general discussion of the structure of comparable membrane proteins. Examples from natural membrane systems have been cited, though a great deal of work on the lipid-water systems (myelin figures), black lipid membranes, and reconstituted membranes provided the fundamental basis for current investigations.

With a view toward limiting the list of references, I have often resorted to citation of the most recent publications and reviews by authors whose previous work is quoted. Therefore, such references have been indicated by "reviewed by or see . . ." in parentheses in the text.

S. K. MALHOTRA

Edmonton, Alberta
June 1983

ACKNOWLEDGMENTS

All the electron micrographs included in this monograph have been produced by myself or one of the associates in my laboratory. I am grateful for the invaluable help I have received from several members of my laboratory, namely Dr. J. P. Tewari, Mr. S. S. Sikerwar, Mr. Steve Ross, Mrs. M. Wong, Miss H. Good, and Miss R. Frei. Mrs. D. Arbuthnott's perseverance in typing is greatly appreciated. My sons, Shantanu and Atul, ungrudgingly provided up-to-date computer printouts of the relevant literature. It would be ungracious not to acknowledge here the enthusiastic support from my wife, Kamini.

My research work has been supported by grants awarded by the Natural Sciences and Engineering Research Council of Canada. The president of the University of Alberta kindly provided funds for the preparation of this monograph. I am grateful to several authors who have provided some of the illustrations and prepublication copies of their manuscripts.

I am grateful, also, to the editorial and production staff of John Wiley & Sons for taking pains in editing the manuscript, and its publication.

S.K.M.

CONTENTS

ABBREVIATIONS

Ach	Acetylcholine
AchE	Acetylcholinesterase (acetylcholine hydrolase, EC 3.1.1.7)
α-BGT	α-Bungarotoxin
BTX	Batrachtoxin
BOTX	Botulinum toxin
cAMP	Cyclic adenosine monophosphate
cGMP	Cyclic guanosine monophosphate
CD	Circular dichroism
CNS	Central nervous system
Con A	Concanavalin A
E face	Concave fractured face (exoplasmic half)
ESR	Electron spin resonance
FSH	Follicle-stimulating hormone
GDP; GTP	Guanosine diphosphate; Guanosine triphosphate
IMPs	Intramembranous particles in freeze fracture
IR	Infra red
LDL	Low-density lipoprotein
NAD	Nicotinamide-adenine dinucleotide
NMR	Nuclear magnetic resonance
ORD	Optical rotatory dispersion
P face	Convex fractured face (cytoplasmic half)
PNS	Peripheral nervous system
Protein kinase	(ATP: protein phosphotransferase, EC 2.7.1.37)
rbc	Red blood cells
S values	Sedimentation coefficients

SBA	Soya bean agglutinin
STX	Saxitoxin
TEA	Tetraethylammonium chloride
TTX	Tetrodotoxin
UV	Ultraviolet
WGA	Wheat germ agglutinin

1

Introduction

The plasma membrane is the essential physiological barrier at the cell surface (Nägeli, 1855) and consists of a now well-established lipid bilayer (Gorter and Grendel, 1925) and associated membrane proteins. Some of these proteins are loosely attached to the membrane, whereas others are more strongly intercalated in the bilayer. Both lipid molecules and membrane proteins may carry carbohydrate moieties that in some cell types extend a good deal out from the lipid bilayer and form a well-defined layer recognizable in electron micrographs. On the cytoplasmic side the membrane proteins interact with other fibrous cytoskeletal elements, which are now known to be relevant to the functional organization of membranes. Thus, the concept of a plasma membrane should incorporate such structural features as may exert a direct effect on its functional parameters.

A simplistic view of the plasma membrane would be that it is a conglomeration of functional domains that may interact with each other and thus govern the activities of the entire membrane. Such an approach is implicit in this monograph since certain functional domains have been selected for consideration. It is also recognized that such an approach may give an impression of the existence of discrete regions in the plasma membrane. In some cases such functional regions are clearly identifiable, for example, the synapse or the neuromuscular junction, regions of cell-to-cell contacts, coated pits, nodes of *Ranvier* in myelinated axons, the apical region of the sperm head, and the purple membrane of the halobacterium. In other cases, for instance, red blood cells, the plasma membrane has a uniform structure, though there may be an underlying microheterogeneity in the distribution of functional domains. Such a simple view of a multifunctional membrane can be reconciled with the current widely held view of the mobility of membrane lipids and proteins by either restricting fluidity to within the functional domains or having no fluidity, as with the Ach receptors in the sarcolemma in the region of the neuromuscular junction (Section 7.5) or in the normal adult human rbc (Kehry et al., 1977). However, such a notion of restricted mobility within the functional domains is likely to be weakened when one considers that the sites of exocytosis and endocytosis of synaptic vesicles, for example, appear to be in different regions of the presynaptic endings (Heuser and Reese, 1973). Also, in the biosynthesis of membrane proteins (Section 15.5), the site(s) of insertion into membranes may be different from the site(s) of assembly into functional units. In the absence of an as yet complete understanding of the process of biosynthesis and the interactions between various seemingly apparent functional domains, the molecular organization of plasma membranes remains poorly understood.

The plasma membrane serves various functions in a cell, so many, served by such a variety of cell types in an organism, that the scope of this monograph must be restricted to consideration of a few selected topics. These topics represent the bias of the author, but some of them are areas of recent advances and others are currently being investigated. Although most are taken from animal cells, particularly mammalian, their general features should be universally applicable to biological membranes, and the rationale for their inclusion appears in the respective section of each chapter. The structure of the purple membrane of the halobacterium represents the best-understood membrane protein system and is therefore included here. Emphasis is laid on the correlation of structure with function and, in particular, advances that have been made since the unit membrane concept was proposed by Robertson in the late 1950s (see Robertson, 1960). The unit membrane concept evolved after the introduction of electron microscopy into biological research and includes advances made since the classical Danielli-Davson (Danielli and Davson, 1935) model was published. More recent advances in membrane biology have been in the elucidation of lipid-protein interactions, in the structure of membrane proteins that traverse the lipid bilayer (transmembrane proteins), and in the fluid and highly dynamic nature of the membranes. These conceptual advances have gone hand in hand with technological advances in membrane biology. One technique is freeze-fracture and freeze-etching for electron microscopy (Bullivant, 1974; Steere, 1957), which has provided perhaps the best available evidence in favor of the lipid bilayer structure of biological membranes (Bretscher and Raff, 1975). During freeze-fracturing the membranes split in half, revealing the two internal fractured faces and the fracture plane follows the hydrophobic interior of the membrane (Branton, 1966; Branton et al., 1975). In addition, the true surfaces of the membranes can be also visualized by etching. This provides the means to detect membranous components (e.g., receptors and antigens) that extend from the bilayer by using appropriate ligands (Tipnis and Malhotra, 1979).

High-resolution micrographs of crystalline arrays of membrane proteins, taken at a low dose of electrons to minimize radiation damage, have been exploited to determine the three-dimensional structure by Fourier transform (Henderson and Unwin, 1975; Unwin and Zampighi, 1980). Improvements in isolation procedures by using various kinds of detergents and protein analysis have added a new dimension to the understanding.

Spectroscopy (IR, UV, CD, ORD, NMR, ESR), fluorescence photobleaching, electrophysiology, and immunology have contributed to advances in membrane biology in a large variety of natural and model membrane systems. The models include lipid-water systems (myelin figures; Luzzati and Husson, 1962; Stoeckenius, 1962), lipid bilayers (thin lipid membranes or black membranes; Mueller et al., 1962) and liposomes (see Kell, 1981), and reconstituted natural membranes. Furthermore, the availability of affinity agents and toxins such as α-BGT (Chang and Lee, 1963) and TTX (Narahashi, 1974) has provided valuable tools for combining morphological and physiological investigations. Recent developments in hybridoma technology to produce monoclonal antibodies (Köhler and Milstein, 1975; Milstein, 1981) provide additional valuable means to probe the structure and functions of membranes.

The plasma membrane can be readily recognized by its location in intact cells by electron microscopy, and biochemical analysis of characteristic properties (enzymatic activity, receptors) of particular cells facilitates identification in isolated fractions. However, there appears to be no single universal marker for the plasma membrane (reviewed by Glick and Flowers, 1978). The cholesterol-to-phospholipid ratio is often used for this purpose for animal cells but the reported values range from 0.3 to 1.4. A commonly used marker for the plasma membrane is 5'-nucleotidase activity, but the extent of activity is not always reported. Adenylate cyclase seems universally associated with the plasma membrane but has also been reported in intracellular membranes (Section 11.3). Biochemical characterization of the plasma membrane is not further discussed in this monograph, though an attempt has been made to include such features as can be clearly identified with it.

2

Current View of the Plasma Membrane

2.1. FLUID-MOSAIC MODEL

It is now widely held that the principal features of the molecular organization of all membranes of a cell conform to the fluid-mosaic model of Singer and Nicolson (1972). In this model, the bulk of the phospholipids are arranged in the form of a discontinuous bilayer with their polar heads in contact with water, and the proteins are either bound to the charged surface of the lipid bilayer (mainly by electrostatic interactions) or intercalated to varying degrees into the hydrophobic interior of the bilayer. These latter categories of proteins are arranged in an amphipathic structure, that is, with the polar groups protruding from the membrane into the aqueous phase and the nonpolar groups largely intercalated in the hydrophobic interior of the membrane. A membrane in this model is envisaged as a two-dimensional solution of proteins in a viscous phospholipid bilayer. Both lipids and proteins are capable of lateral mobility in the plane of the membrane (Fig. 2.1).

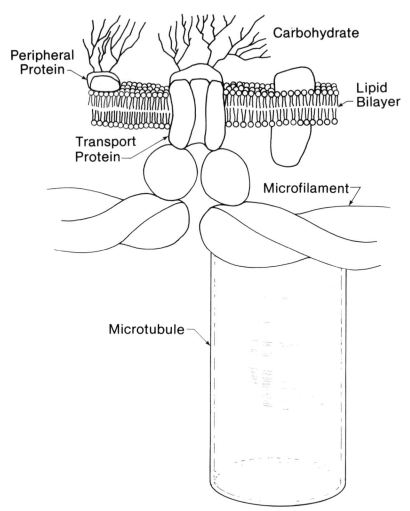

Figure 2.1 A schematic representation of the plasma membrane showing lipid bilayer, a peripheral protein, and a transport protein such as Ach receptor protein, which is an integral membrane glycoprotein. Another integral protein is shown on the right (unlabeled). The transport protein is shown to interact with cytoskeletal elements such as microfilaments and microtubules. (Not to scale)

2.2. LIPID PHASES IN THE PLASMA MEMBRANE

Although the lamellar bilayer is the most prevalent conformation of lipid characterized in natural membranes, conformation other than bilayer, for instance, hexagonal, may occur transiently as during fusion (Chapter 14). Notwithstanding the role of nonlamellar lipid phases in natural membranes (if any), phospholipids are capable of forming complex macromolecular assemblies in lipid water systems (Glauert, 1967; Lin et al., 1982). For example, a mixture of cardiolipin and phosphatidylcholine show liposomes and long tubes. These structures may be multilamellar or unilamellar. In the presence of Ca^{2+}, helical liposomes can be observed by light microscopy. Ca^{2+} concentrations as low as 10^{-6} M produced more helices than higher concentrations. At higher concentrations of Ca^{2+}, more complex structures appeared, presumably as a result of the collapse of helices. The helix formation is thought to represent membrane-to-membrane attachments mediated by Ca^{2+} (see Lin et al., 1982). It appears that other phospholipids may also form tubular and helical macromolecular assemblies, for instance, egg lecithin (Bangham quoted in Lin et al., 1982). Ca^{2+} is thought to be involved in membrane fusion, and its presence facilitates lateral phase separation of acidic phospholipids (Lin et al., 1982). Acidic phospholipids (such as cardiolipin, phosphatidylserine) which prefer bilayer organization tend to adapt a hexagonal phase in the presence of Ca^{2+}.

Irrespective of their physical conformation, lipids perform essential roles in natural membranes. What specific lipids are required in ion transport proteins and membrane-bound enzymes is a matter of current interest. For example, rat brain adenylate cyclase seems to require lysophosphatidylcholine or sphingomyelin for catalytic activity, and the cell may contain some inhibitory lipid that can be exchanged with a stimulatory lipid by phospholipid exchange protein. It can also produce lysophosphatidylcholine by hydrolysis of phosphatidylcholine (Hebdon et al., 1982). Attempts are being made with a number of membrane proteins to reconstitute their functional systems to understand the role of specific lipid-protein interactions (Kracke et al., 1981; Montal et al., 1981). The specific requirement appears to be in the polar head groups, rather than in the fatty acid chains (Sandermann, 1978).

It appears from recent data derived from deuterium NMR studies that enable measurements on a time scale of 10^{-4} seconds (as compared to 10^{-8} seconds with ESR) that there is no long-lived lipid shell surrounding integral membrane proteins (see Chapman et al., 1982). However, in most membranes, the phospholipid–protein ratios are such that most of the lipid is likely to be perturbed by protein, even that lipid which is removed by several solvation layers from the one in immediate contact with the surface of the protein. The lipid in contact with the protein has been termed the boundary lipid (Longmuir et al., 1977; Warren et al., 1975). Some very small amounts of lipid could be present in the form of clusters, as demonstrated for phosphatidylethanolamine in rbc (Marinetti and Crain, 1978). But these domains in rbc appear to be homogeneous in respect of lipid organization, as indicated by a single decay time of a fluorescent-labeled hydrophobic probe (Karnovsky et al., 1982). However, such a free lipid is likely to be a very small fraction of most natural membranes. This lipid may not play any important biochemical role, except in the myelin sheath that has 70–80% of its contents as lipid, in membranes undergoing fusion when protein-free lipid domains are created (Chapter 14), and during growth of membranes when lipid vesicles, without incorporated protein, are inserted into the membrane (Chapter 15).

The observations that there are distinct physical lipid domains in many membranes (Karnovsky et al., 1982) appear to be in conflict with the conclusion that all the phospholipid of a membrane is in a homogeneous phase on a time scale of 10^{-4} seconds (see above). Further studies should clarify whether there are differences in lipid phases in different membranes or whether different phases exist in the same membrane. However, specific lipids are likely to be associated with particular membrane proteins, thus creating functional domains. Lipid domains could be created artificially when crystallization of the lipid chains occurs below the phase transition temperature, which would lead to a phase separation of high protein concentrations and lipid patches, the latter corresponding to pure lipid (Chapman et al., 1982). Such a phenomenon is visualizable in

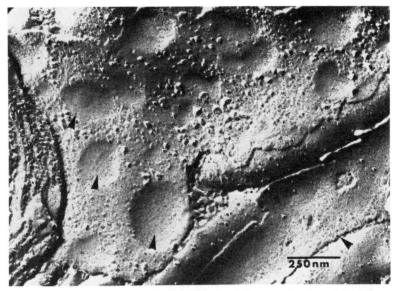

Figure 2.2   Freeze-fracture preparation of the rat sciatic nerve showing relatively large areas of the fractured plasma membrane devoid of IMPs (arrows). Such an appearance can result from lateral phase separation of lipids during freezing; when the freezing rate is not high, the protein can be excluded from lipid patches prior to being frozen. (Kleemann and McConnell, 1974)

freeze-fracture replicas when the cooling, presumably, is not rapid (Fig. 2.2).

Further studies on the reconstitution of membranes, in particular, should advance our current understanding of the role of specific lipids in functional domains and the lipid-protein interactions. Proteins disorder the acyl chains of the lipids and restrict their mobility, but the precise functional implications of lipids at the protein interface are uncertain (Watts, 1981).

2.3. CARBOHYDRATE SURFACE COAT

A number of membrane proteins and lipids carry carbohydrate side chains. Such plasma membrane glycoproteins and glycolipids serve as surface receptors for specific ligands, for instance, hormones, antigens, lectins, thrombin, Ach, virus, and antibodies (reviewed in Hughes, 1976; Lackie, 1980). Cell surface glycoproteins serve in the adhesion of cells, as, for example, in the adherence of blood platelets to collagen in haemostasis. Interactions among complementary structures—glycosyl transferases and oligosaccharide acceptors—could play a vital role in cell adhesion (Roth, 1973). Glycosyl transferases may also serve in the formation of intercellular junctions (desmosomes and tight junctions) by repeated reaction between glycosyl transferases and their substrates on the adjacent cells.

Normally, the basic protein of the myelin sheath does not carry carbohydrates, but threonine residue is glycosylated by a N-acetylgalactosaminyl transferase from submaxillary glands. Possibly the myelin basic protein undergoes glycosylation transiently during its biosynthesis and insertion into the myelin sheath. The role of this carbohydrate moiety in myelination remains to be investigated (see Hughes, 1976).

Glycoproteins provide a great deal of specific interactions because they can exist in several different forms. For example, a disaccharide (galactosyl-N-acetylglucosamine) can have 14 different forms at the terminal residues of a glycoprotein. (In contrast, a dipeptide containing L-amino acids can have only two forms [see Hughes, 1976].) Also, glycolipids occur in many different forms; at least 37 different glycosphingolipids were detected in the rat small-intestinal cells with distinct differences in the carbohydrate moieties in the epithelial (endodermal origin) and non-epithelial (mesodermal origin, with the exception of nerve cells) cells. Some of these carbohydrate moieties representing cell surface antigens may appear and disappear at different developmental stages and thus determine the developmental sequence (Breimer et al., 1982).

Carbohydrates of the plasma membrane glycoproteins and glycolipids extend to a varying extent into the extracellular material, and may form a prominent carbohydrate-rich layer on the surface of the cells. This layer has been referred to as the cell coat, glycocalyx, or fuzzy layer, and, together with the lipid bilayer, as the greater membrane (see Schmitt and Samson, 1969). Though it may be poorly defined morphologically on many cell surfaces, its presence has been demonstrated in a large variety of cells by using histochemical approaches such as staining with colloidal thorium or periodic acid-silver (a modification of the well-known periodic acid-Schiff) which specifically detect carbohydrate moieties (Rambourg and Leblond, 1967). It is assumed that a carbohydrate cell coat is a ubiquitous feature that may not be very conspicuous in routinely prepared thin sections for electron microscopy.

In the intestinal epithelial cells, the microvilli show such a carbohydrate-rich layer, ~100 nm thick, that is composed of fine filaments radiating from the plasma membrane. It is highly resistant to lytic agents (EDTA, Ringer, neuraminidase, hyaluronidase, trypsin, Papsin), and treatments short of membrane dissolution were ineffective in removal of the filamentous coat from the plasma membrane. Therefore, this filamentous outer layer should be considered an integral component of the plasma membrane and not an extraneous material loosely associated with it (Ito, 1965). In the amoeba *Choas choas,* the outermost filamentous layer consists of fairly coarse filaments distributed over the entire cell surface (see Schmitt and Samson, 1969).

The carbohydrate cell surface components of the plasma membrane may interact with extracellular constituents such as fibronectins and basal lamina (basement membrane that forms a morphologically continuous layer in normal organized tissues [Fig. 5.1]). An example of the role of basal lamina in the neuromuscular junction is discussed in Chapter 8. On the cytoplasmic side proteins of the plasma membrane may interact with the skeletal elements (Chapter 12), and various functional domains may interact with each other to regulate and govern the functioning of the entire plasma membrane.

2.4. ASYMMETRY OF THE PLASMA MEMBRANE

Consistent with the functional differences between the extracellular and cytoplasmic sides of the plasma membrane, the asymmetric distribution of the lipid and protein on the two layers of the membrane is well recognized (reviewed in Bretscher and Raff, 1975; Op den Kamp, 1979), despite differences in the results reported on the distribution of specific amounts of cholesterol in the two layers of the plasma membrane (Blau and Bittman, 1978). In general, it can be stated that there appears to be no absolute asymmetry in the distribution of lipids; that is, each phospholipid is present on both sides of the membrane though in different amounts. The protein asymmetry appears to be absolute in that there is no known protein that is exposed on both surfaces of the plasma membrane and that has a similar structure on either surface. Carbohydrates, whether attached to the protein (glycoproteins) or lipids (glycolipids), are exposed at the extracellular surface of the plasma membrane.

The asymmetric nature of the plasma membrane is further strengthened when one considers the cytoskeletal elements on the cytoplasmic side and extracellular constituents such as fibronectins, basal lamina on the external surface that interact with the plasma membrane and exercise regulatory control on the functional aspects of the plasma membrane. The asymmetry of the plasma membrane is best illustrated in the complementary freeze-fractured faces (P and E) that generally show more intramembranous particles (IMPs) on the P face than on the E face. These IMPs are mostly representations of integral membrane proteins, though the possibility remains that lipid may also contribute to the appearance of IMPs in freeze-fractured faces in certain cases (Chapter 3; Malhotra et al., 1981).

The chemical asymmetry of the plasma membrane has been demonstrated by using permeant and nonpermeant probes, lectins, specific ligands, and specific histochemical staining procedures. Toxins such as TTX and TEA have been effectively utilized in the demonstration of asymmetric organization of sodium and potassium channels in the excitable membranes.

Despite the recognized ability of the lipid molecules to diffuse laterally and to undergo flip-flop motion (Section 2.5), the asymmetric distribution of lipid molecules in the bilayer in the plasma membranes are apparently retained. It would thus seem that diffusion of lipid molecules is restricted to lateral diffusion in natural membranes unless biosynthetic events involve flip-flop motion. Lateral diffusion would be restricted to within the functional domains.

Flip-flop of membrane proteins is not known to occur (see Stryer, 1981), and therefore protein asymmetric distribution is retained once established biosynthetically. The effect on the asymmetry by vertical displacement of protein molecules (e.g., rhodopsin during photobleaching; Mason et al., 1974) is not known.

2.5. FLUIDITY

The experimental basis and the background for the fluid-mosaic model emerged gradually during the last two decades. In one of the earlier publications supporting the fluid nature of the membrane constituents, Frye and Edidin (1970) noted that cell surface antigens were seen to undergo rapid intermixing after formation of mouse-human heterokaryons, and the plasma membrane was considered to be fluid enough to allow free diffusion of surface antigens. In view of the diversity in the chemical composition and interactions among molecular constituents of the membranes, the extent of the fluid contents may differ, not only in different plasma membranes but in different functional domains in an otherwise continuous membrane. The fluidity may undergo changes with development and growth and also undergo alterations with physiological phenomena (de Latt et al., 1980; reviewed in Malhotra, 1980). It is also regulated by cytoskeletal elements by interacting with membrane proteins (see Smith and Palek, 1982).

In general, the lateral movement of lipids is faster than that of the proteins. For example, Schlessinger et al. (1977) determined the diffusion constant of a fluorescent lipid probe to be $\sim 9 \times 10^{-9}$ cm²/s and that of the protein to be $\sim 2 \times 10^{-10}$ cm²/s in the plasma membrane of L-6 myoblasts. The lipid probe was free to diffuse over long distances (more than 4 μm). However, the extent of free lipid capable of diffusion in different membranes and in different functional domains of a membrane may vary a great deal. The diffusion of such a lipid may also be restricted by the lipid-protein interactions. Also, the diffusion of protein may be highly restricted, to the extent that they essentially appear immobile.

In addition to lateral movement, lipid molecules are also capable of flip-flop (tumbling from one half of the bilayer to the other). Not much is known about the rates at which lipid molecules undergo this type of movement in natural membranes. There is no evidence that any known natural membrane protein undergoes flip-flop movement (Stryer, 1981). However, proteins are capable of rotational diffusion in the hydrophobic core of the membrane and perpendicular to the membrane plane (see Cone, 1972; Hoffmann et al., 1980). Such a movement of proteins could be facilitated by changes in the fluidity of membrane lipid, for example, by changes in the cholesterol contents (Borochov and Shinitzky, 1976). That such a displacement of protein may take place in natural membranes as a functional modulation is suggested by the results of photobleaching of the isolated retinal disk membranes: A sinking of the intramembranous particles, presumably photopigment molecules, from the polar surface was reported in freeze-etching experiments (Mason et al., 1974). A somewhat similar effect of dark adaptation on the intramembranous particles in the plasma membrane of the growing zone of sporangiophore of the fungus, *Phycomyces,* which is phototropic, was observed (Section 5.8; Tu and Malhotra, 1975). However, these initial studies require confirmation by using suitable surface markers for proteins.

2.6. REGULATION OF MEMBRANE FLUIDITY

In prokaryotes, other than Mycoplasmas, membrane fluidity is regulated by varying the number of double bonds and the chain length of their fatty acids (Chapman et al., 1967; see Stryer, 1981). Mycoplasmas, the wall-less prokaryotes, are the only prokaryotes that require cholesterol for their growth and can incorporate cholesterol from the culture medium (Rottem, 1981). Cholesterol appears to maintain an intermediate fluid state by eliminating phase transition at low temperatures and decreasing membrane fluidity at high temperatures (Chapman, 1980). Several species of Mycoplasmas preferentially incorporate saturated fatty acids from the growth medium, and therefore the incorporation of cholesterol into their plasma membrane prevents crystallization of lipids at their optimal growth temperature (Rottem, 1981). Mycoplasmas represent perhaps the best-known case among natural membranes in which the role of cholesterol as a regulator of membrane fluidity has been elucidated.

Cholesterol is the best regulator of membrane fluidity in eukaryotes also (see Stryer, 1981). The availability of mutant cell lines defective in cholesterol biosynthesis made it feasible to investigate the influence of cholesterol on membrane properties.

Sinesky et al. (1979) demonstrated that the physical state of the membrane lipid, that is, the order parameter, influences the rate of catalysis of $(Na^+ + K^+)$-ATPase. A somatic cell mutant of the Chinese hamster ovary cell line (CHO-K1) defective in cholesterol biosynthesis could be grown under defined conditions with various amounts of cholesterol. The enzyme activity was found to vary by a factor of 10 as the order parameter was varied. By comparison with the plasma membrane of the parent cell line, which had different cholesterol contents but a similar acyl chain order parameter, it was shown that the cholesterol has a physical rather than a chemical effect on the enzyme activity. This finding suggests that some other membrane-bound enzymes could also be influenced in a similar manner by alterations in the acyl chain order parameter. For example, drug treatments, hormone treatments, and changes in membrane lipid could influence the acyl chain parameter.

Alterations in membrane fluidity may result from phospholipid methylation (phosphatidylethanolamine to phosphatidylcholine) which has been reported to be involved in signal transduction (Axelrod, 1982). Methylation of phospholipids could be related in sequential events that result in cellular response of receptor-mediated stimulation of leukocytes, T-lymphocytes, and mast cells (see Bareis et al., 1982): For instance, stimulation of rabbit neutrophils by the chemotactic peptide f-Met-Leu-Phe produces release of a fatty acid, arachnidonic acid, by methylation of phospholipids. Release of this fatty acid requires activation of membrane-bound phospholipase A2, which is dependent upon extracellular calcium influx. The released arachidonic acid can be remetabolized by the cell. Methylation of phospholipid may be related to the increased fluidity of the membrane and/or a change in the lipid environment around the calcium channel (Bareis et al., 1982).

3

Intramembranous Particles

One of the major advances in the elucidation of membrane structure occurred with the development of freeze-fracturing-etching techniques for preparation of biological materials for electron microscopy (Steere, 1957). When biological membranes are prepared by freeze-fracturing, they split along their hydrophobic mid-planes at low temperatures (Branton, 1966). This technique has provided some of the strongest available evidence that all biological membranes are based on a lipid bilayer. It is the only technique now available that makes possible a direct examination of the interiors of membranes. It also enables one to view the true surface (etched-face) of the membrane by etching, and surface molecules can be labeled by appropriate ligands and tracers (see Tillack and Marchesi, 1970; Tipnis and Malhotra, 1979). Freeze-fracturing can also be combined with cytochemical studies (Pinto da Silva et al., 1981). When the freeze-fractured faces of plasma membranes, suitably replicated by evaporation of carbon-platinum or other suitable metals, are examined by transmission electron microscopy, they reveal a varying number of intramembranous particles (IMPs) on the two complementary faces. There are usually more IMPs on the face that is apposed to the cytoplasm (P face) than on the face that is apposed to the outside (E face, Branton et al., 1975) so that the two fractured faces appear asymmetric. By using tracers (e.g., virus, and Con A or wheat germ agglutinin labeled with ferritin molecules), it has been demonstrated that some of the IMPs are associated with the receptors that are exposed on the surface of the erythrocyte plasma membrane (Marchesi et al., 1973; Pinto da Silva et al., 1981). Apart from the red blood cell plasma membrane, several other plasma membranes have been shown convincingly to contain IMPs that represent transmembrane proteins. Some of the well-characterized examples are bacteriorhodopsin in the purple membranes of halobacteria (Chapter 9), acetylcholine (Ach) receptors in the electric organ of the electric fish, the postsynaptic membrane in the vertebrate neuromuscular junction (Chapter 7), and connexin in the cell contacts that are involved in intercellular communication (gap junctions, Section 6.3).

Based on reconstitution experiments in particular, it is now commonly accepted that the IMPs represent integral membrane proteins in natural membranes (reviewed by Malhotra, 1980). In the absence of proteins, the fractured faces often appear smooth and devoid of IMPs. However, IMPs have been seen in freeze-fractured faces of artificial membranes made from lipid alone (Verkleij et al., 1979; Sen et al., 1981). Moreover, Robertson has raised the possibility that some of the IMPs may be lipid in nature (Robertson et al., 1980) and arise artificially during preparation (Robertson and Vergara, 1980). Such a possibility should be borne in mind in the critical evaluation of the significance of IMPs. Artifacts could arise from condensation, plastic deformation, and metal decoration. The IMPs may vary in size from ~8 to 15 nm in diameter or even larger (Tipnis and Malhotra, 1976), though in most plasma membranes they are reported to be around ~8 to 10 nm. Unless variation in size is an artifact because of variations in metal thickness or temperature variation during freeze-fracturing, their protein content would seem to differ. In one case the protein contents of the IMPs representing bacteriorhodopsin molecules have been estimated (Chapter 9). The pattern of distribution of IMPs can be altered experimentally by manipulation of, for example, pH (Yu and Branton, 1976), cAMP (Tu and Malhotra, 1977), Ca^{2+} (Peracchia, 1980), and glutaraldehyde (Sikerwar and Malhotra, 1981). Their distribution has been reported to change during fusion of plasma membranes because of redistribution of integral membrane protein. Entry of malarial parasite into red blood cells is accompanied by changes in the distribution of IMPs (Aikawa et al., 1981). The pattern of distribution of IMPs can facilitate identification of morphological and presumably functional regions in otherwise continuous plasma membrane such as in epithelial cells, various processes of neurons, and in single-cell aerial hyphae of *Phycomyces* which have a distinct photoreceptive zone (Bergman et al., 1969; Section 5.8). It is therefore tempting to question the precise significance of the distribution of IMPs in relation to their role in functions associated with the plasma membrane.

Gingell (1976) has proposed that the IMPs that are glycoproteins partially embedded in the lipid bilayer are in a dispersed state because they suffer mutual electrostatic repulsion caused by sialic acid charges. "A reduction in pH sufficient to reduce the acidic charge or an increase in ionic strength which lowers the electrostatic potential of the charges" will cause the IMPs to attract one another by Van der Waals forces. This attraction between the IMPs could be strong or weak, and in the latter case it is reversible. Gingell has speculated that the factors that influence the electric repulsion may control the extent of aggregation of IMPs and thus provide a mechanism for permeability changes. The inference of this suggestion is that diverse cellular functions are due to changes in membrane permeability that are accompanied by aggregation of IMPs. Gingell has discussed evidence from the available literature to elaborate his hypothesis. His examples include changes in the plasma membrane during pinocytosis induced by cationic substances, changes in ion permeability of *Xenopus* egg membrane induced by polycations (polylysine, polyarginine), patching and capping response of lymphocytes to antibodies, and aggregation of IMPs in gap junctions. (The structure of gap junctions is given in Section 6.3.) It should be interesting to test Gingell's model and to know whether aggregation of IMPs is necessarily related to changes in membrane permeability or vice versa. However, the mechanism of spontaneous aggregation of

IMPs in biological membranes, for example in the formation of gap junctions or their clustering in the postsynaptic membrane in neuromuscular junction, does not seem to be understood. (Once aggregated, adherence to the cytoskeletal elements could keep the IMPs from diffusion.) Following a suggestion by Dr. Israel Miller, Gingell has considered that aggregation of IMPs might be brought about by a phenomenon comparable to concentration-dependent colloid flocculation.

4

General Remarks on Membrane Proteins

4.1. INTRODUCTION

The various functional domains of a plasma membrane are recognizable by their characteristic membrane proteins. Lipids associated with such domains provide an appropriate environment for the action of proteins (Stryer, 1981). (But there may not be a long-lived phospholipid shell surrounding proteins; more recent studies by NMR on a time scale of $\sim 10^{-4}$ seconds [vs. ESR 10^{-8} seconds] indicate that all the phospholipid is in a homogeneous phase [see Chapman et al., 1982].)

Operationally, membrane proteins are regarded as peripheral (extrinsic) or integral (intrinsic), depending on the ease with which they can be dissociated from the membrane (Capaldi, 1977; Singer and Nicolson, 1972). Peripheral proteins usually require mild treatment, such as high ionic strength of solution, or a change in pH, to release them from the membrane, and are thus likely to be associated with membranes by weak noncovalent (mostly electrostatic) forces. Integral proteins interact with the hydrocarbon chains of membrane proteins and require surfactants (detergents and bile salts) to remove them from the membrane. Even after removal from the membrane, they often remain associated with lipid which is required for their biological function.

Acetylcholinesterase (AchE, Chapter 8) and 43,000 molecular weight (MW) protein associated with the Ach receptors (Section 7.7) are the two examples of the peripheral proteins considered in this monograph. Other examples include spectrin (Chapter 12) and actin in the rbc and several cytoskeletal proteins that interact with the plasma membrane in various eukaryotic cells. Bacteriorhodopsin (Chapter 9) of the purple membrane in halobacteria, nicotinic acetylcholine (Ach) receptor (Chapter 7), porin (Section 6.4), and glycophorin are examples of some of the integral membrane proteins discussed at length in this volume. Other examples in this category include Band 3 protein in the rbc plasma membrane, adenylate cyclase, Ca-ATPase, (Na^+, K^+)-ATPase, rhodopsin, connexin of the gap junctions, lipophilin of the myelin sheath, and ion transport channel proteins (Chapter 10). Such proteins are thought to span the bilayer exposed at both surfaces of the lipid bilayer (see Boggs and Moscarello, 1978). Most peripheral proteins are bound to the integral membrane proteins (Stryer, 1981).

The criteria used in classifying membrane proteins are not very clearly defined since some of the proteins may fall into both categories. Some of the membrane proteins such ATPase complex in the inner mitochondrial membrane and cytochrome b5 in the electron transport chain of the endoplasmic reticulum have components that meet the criteria for both peripheral as well as integral membrane proteins (Dehlinger et al., 1974; Racker, 1976).

Apart from the myelin sheath, which has a relatively low protein content (~20%), most animal cell plasma membranes contain nearly 50% protein, whereas membranes of bacteria contain 80–90% protein (see Finean et al., 1978). While these estimates are for overall membrane protein contents, the specific proteins are likely to differ in different functional domains. It is of current interest to determine if integral membrane proteins have common structural features, an issue summarized in the following section.

4.2. INTEGRAL MEMBRANE PROTEINS

The number of possible conformations that can be adapted by the integral membrane proteins is restricted by the specialized environment of the lipid bilayer. Usually a typical soluble protein consists of regions of a regular secondary structure (α-helices and β-pleated sheets) connected by a nonregular structure. Regular secondary structures are energetically stabilized by the internal hydrogen bonds between peptide amide and carbonyl groups. Whenever the dihedral angles (ϕ and ψ) of the backbone of a polypeptide chain are periodically repeated, a helix (linear group) is generated, whereas the β-pleated sheets result from the hydrogen bonding among the extended polypeptide chains (Ramachandran and Sasisekharan, 1968). Since the regular secondary structures can satisfy the maximum hydrogen bonding requirements internally, they may well exist within the hydrophobic region of the lipid bilayer. Three types of known regular secondary structures exist as energetically stable configurations of the intramembranous section of integral membrane proteins, namely, α-helices, β-sheets, and the third conformation, which has been proposed as an intramembranous part of integral membrane protein, called β-helix (Kennedy, 1978). In the β-helices the individual amino acid residues could be nonequivalent (L and D-amino acids) and the amino acid side-chains are all oriented in the helix interior, which thus makes them potential candidates for ion transport purposes. Since the presence of the hydrophobic region in the lipid bilayer calls for the hydrophobic amino acid side-chains to be in direct contact with the alkane portion of the bilayer, an analysis of the sequences of the integral membrane proteins should lend important clues to its secondary organization in the membrane interior. Prediction of the local secondary structure from the amino acid sequence data alone is now feasible, with limited success, for the soluble proteins (Chou and Fasman, 1978).

In the water-soluble proteins the interior houses hydrophobic residues, whereas the surface is predominantly comprised of hydrophilic side-chains exposed to the solvent water. Rose (1978) has developed a method for the detection of hydrophobicity distribution over a protein sequence (Rose and Roy, 1980). This has been claimed to discern the hydrophobic regions of the interior sequence from the hydrophilic regions of the exterior sequence. Based upon this scheme, Kyte and Doolittle (1982) described a computer program that examines objectively the hydrophilic and hydrophobic natures of a polypeptide chain by employing a *hydropathy* scale which includes both the hydrophilic and hydrophobic tendencies of each of the 20 amino acid side-chains. The intramembranous segments of a number of integral membrane proteins can be identified with this method. Since only a few of the integral membrane proteins have thus far been successfully sequenced, any generalization with regard to the prediction of the secondary structure should be cautious. The details of the structure of the bacteriorhodopsin, the best understood integral membrane protein, and glycophorin, one of the two major membrane proteins of rbc, are given in Chapter 9 and Section 4.3 respectively.

The mechanism(s) of the insertion of integral membrane proteins into lipid bilayers is discussed in Section 15.4.

4.3. GLYCOPHORIN

Tomita and Marchesi (1975) have reported on the amino acid sequence of a glycophorin (MW 31,400) which is one of the two major transmembrane proteins in the human erythrocytes, the other being Band 3 protein (MW 90,000; Bretscher, 1971). Glycophorin has three distinct domains—a glycosylated segment, consisting of 64 amino acids to which 16 oligosaccharide chains are attached, a COOH terminal consisting of 35 amino acids, and a middle hydrophobic segment consisting of 32 amino acid residues. Glycophorin carries NM blood group system (Furthmayr, 1978); it also carries influenza virus binding sites and phytohemagglutinin and wheat germ agglutinin binding sites (Tomita and Marchesi, 1975) and is a suspect in serving as a receptor for the malarial parasite *Plasmodium falciparum* (Pasvol et al., 1982).

Freeze-fracturing, combined with labeling with wheat germ agglutinin (localized by staining with colloidal gold), shows that the glycophorin may preferentially localize in the outer half of the plasma membrane. This contrasts with the distribution of Band 3 protein (detected by labeling with Con A) in the inner half of the plasma membrane (Pinto da Silva and Torrisi, 1982). Such a distribution of membrane proteins requires confirmation because Fisher (1982) found that the label ^{125}I-FITC-Con A detectable by autoradiography was exclusively confined to the extracellular half of the plasma membrane in freeze-fractured monolayers.

An investigation of the amino acid sequence of glycophorin led Kennedy (1978) to propose a β^{13}-helical structure (the superscript [13] refers to the number of residues per turn) for the membrane intercalated segment. Residues 55–107 have been arranged into four turns of β^{13}-helix which has a hydrophobic external surface: Negative-charged residues are in the interior, which has implications for functioning as a cation channel.

It should be noted that genetic variants of glycophorin molecules exist and carry M or N. The absence of one or the other glycophorin does not appear to alter the structure of function of the rbc plasma membrane (reviewed in Furthmayr, 1978).

5

Functional Domains

5.1. INTRODUCTION

Several instances, some of them obvious, of functional domains in the plasma membrane of a variety of cells, excitable and nonexcitable, are given in the following sections. Most of these domains have been demonstrated by using suitable techniques at the electron microscope level. Other seemingly homogeneous functional domains discussed in this monograph include the intercellular junctions (tight junctions, gap junctions, desmosomes, septate junctions) (Fig. 5.1); junctional sarcolemma in the region of the neuromuscular junction which contains two principal functional systems, namely, Ach receptors (Chapter 7) and AchE (Chapter 8); the purple membrane containing bacteriorhodopsin (Chapter 9); and the coated pits, which internalize selective extracellular materials, for example, low-density lipoprotein (LDL) particles (Chapter 14).

Even within a seemingly homogeneous functional membrane domain, microheterogeneity is likely to exist. On-going investigations into the interactions between such microheterogeneous domains re-

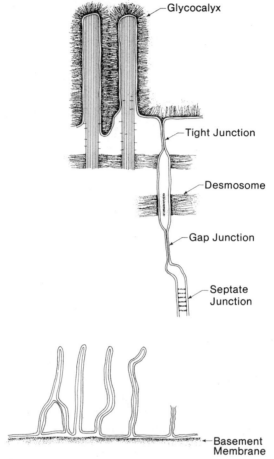

Figure 5.1   A diagrammatic representation of the plasma membrane in an epithelial cell showing the apical microvilli, lateral junctional and nonjunctional domains, and the basal infoldings. The microvilli contain cytoskeletal elements that also interact with the lateral plasma membrane.

veal, for example, that the presynaptic membrane contains calcium channels and sites for fusion of synaptic vesicles for release of the transmitter; also, components that regulate the opening and closing of the channels and the recycling of the membranes of the synaptic vesicles. Demonstration of such domains is becoming feasible as developments in techniques take place.

On the assumption that most of the lipid in natural membranes is associated with proteins, the demonstration of lipid domains would imply the existence of functional domains (Finean et al., 1978). Karnovsky et al. (1982) used 1,6-diphenyl-1,3,5-hexatriene (DPH), a fluorescent probe, which locates itself in the hydrophobic acyl region of the membrane and which could be followed by its decay characteristics. With the exception of rbc, various natural membranes investigated manifested a heterogeneity in their lipid structure. The existence of lipid domains in membranes was further supported by the observation that incorporation of unsaturated free fatty acids into biological membranes did not change the diffusion coefficients, though the gel regions became disordered, thereby indicating that the fatty acid pertubation was local (reviewed by Karnovsky et al., 1982). In the Ach receptor-rich electric organ of the electric fish, *Torpedo,* cholesterol was found to be lacking from the Ach receptor-rich region of the postsynaptic membrane (Perrelet et al., 1982). In the cultured hepatocytes, cholesterol was found to be deficient in the plasma membranes in the region of the adjoining hepatocytes, whereas the free plasma membrane contained a lot more cholesterol (Robenek et al., 1982). Both these studies are based on freeze-fracture data involving the use of filipin, a polyene antibiotic that complexes with cholesterol.

With the availability of monoclonal antibodies that can be used for the detection of antigenic molecules at the electron microscopic level, our comprehension of the functional domains in membranes is likely to advance rapidly.

5.2. PLASMA MEMBRANES OF EPITHELIAL CELLS

Apart from the regional, structural, and functional characterization within the lateral plasma membrane, such as junctional and nonjunctional zones and various types of junctions (Chapter 6), the free surfaces of cells in epithelia show morphological and physiological differences (see Fawcett, 1981). For example, the lateral and basal surfaces of the avian salt gland cells show (Na^+-K^+)-ATPase activity which is lacking from their luminal surfaces. This (Na^+-K^+)-ATPase is ouabain sensitive, and the differential activity could be demonstrated cytochemically as well as autoradiographically by using a specific inhibitor ^{3}H-ouabain (Ernst and Mills, 1977). Such a compartmentalization in the basolateral localization of (Na^+-K^+)-ATPase thought to be related to the secretion by the salt glands, may be common to all transporting epithelia (Ernst and Mills, 1977).

The plasma membrane of the epithelial cells that faces the lumen of the intestine contains sucrase-isomaltase in its active form, an enzyme that has not been detected on the basolateral surface of the cells. Sucrase (MW 140,000) and isomaltase (MW 160,000) are two tightly linked polypeptide chains and anchored to the plasma membrane by a hydrophobic NH_2-terminal segment of the isomaltase subunit (Brunner et al., 1979; Lodish et al., 1981). An immunoreactive precursor of this enzyme is found in the Golgi complex and in the basolateral plasma membrane. This precursor has twice the MW of the enzyme on the apical surface and is presumably converted to the active form by proteolytic cleavage (Hauri et al., 1979; see Lodish et al., 1981). Orci et al. (1981) have reported as a result of their freeze-fracture studies that the basal region of the foot processes of the podocytes differs from the cell body in the number of filipin–sterol complexes and intramembranous particles. The foot processes of the neighboring podocytes interdigitate and rest on the glomerular basal lamina. These podocytes form part of the filtration barrier in the glomerulus.

Virus of different types seem to bud off from different surfaces of the same cell; some from the apical surface, some from the basal, and others from the lateral surface. There is apparently specificity in the interaction between the virus and the molecular organization of the cell that determines the viral bud site (Boulan and Sabatini, 1978).

5.3. COATED PITS

The plasma membrane of many eukaryotic cells contains morphological domains that are referred to as *the coated pits* because of their association with a bristle coat on the outer surface of the lipid bilayer membrane (Fig. 5.2). These bristles (~16 nm) radiate into the

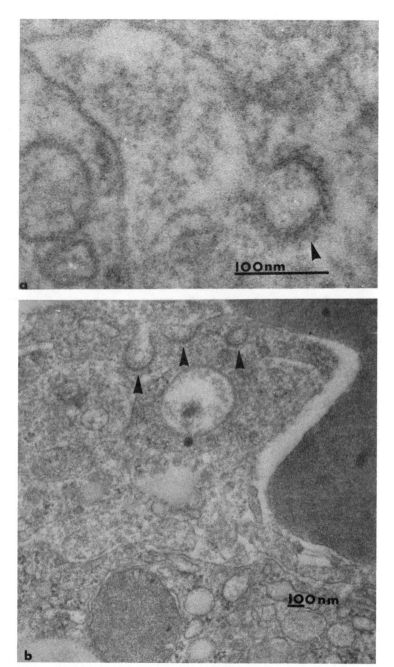

Figure 5.2 Coated pits (arrows) in the plasma membrane of the mouse liver that can be readily recognized in electron micrograph by their dense staining due to their "coat".

cytoplasm (Woodward and Roth, 1978). The coated pits are implicated in the selective internalization (micropinocytosis) of extracellular proteins and form coated vesicles. These serve in the recycling of cell surface receptors, for instance, LDL receptors (Chapter 14) and other membrane components, as in the possible recycling of the membrane of the synaptic vesicles after release of the transmitter by exocytosis (Heuser and Reese, 1973). Coated vesicles are also possibly involved in the intracellular transport and shuttling of various membranes and membrane-bound components (reviewed by Pearse, 1980).

Coated vesicles ($\sim$50–250 nm in diameter) have been isolated from a variety of cells (chicken oocytes, human placenta, pig brain, adrenal medulla, nonsecreting) and show a characteristic lattice of hexagons and pentagons in negatively stained electron micrographs. This lattice, which makes up the coat of the vesicles, contains one major polypeptide, referred to as clathrin, MW 180,000, and two other proteins, MW 100,000–125,000 and 35,000–55,000, in smaller amounts than clathrin (Pearse, 1980; Woodward and Roth, 1978). The coats appear to be easily solubilized from the membrane vesicles by treatment with urea or Mg^{2+}. Removal of these agents leads to reassembly of the coats. Such a flexible capability of the clathrin coats would seem to be well suited for cycling of the coated vesicles from one membrane compartment to another.

Clathrin has been compared to actin in its conservation of amino acid sequence in cells examined from several different animals (Pearse, 1980).

A further discussion of the role of coated pits in the internalization of extracellular proteins appears under "Exocytosis, Endocytosis, and Fusion of Membranes" in Chapter 14.

5.4. PLASMA MEMBRANE OF MAMMALIAN SPERM

The plasma membrane of the mammalian sperm cell provides an excellent example of regional differences in its structure and function (see Fawcett, 1981). Three distinct regions are evident: acrosomal, middle (mid-piece), and tail-piece (principal). These regions can be easily recognized from their characteristic distribution and the size of IMPs. The acrosomal region could be further differentiated into the cap region, which is fusigenic, and the nonfusigenic post-acrosomal region. This differentiation, which was discerned by using cytochemical techniques combined with freeze-fracture electron microscopy, indicated that the fusigenic plasma membrane is rich in anionic lipid, whereas the nonfusigenic membrane is not. These two regions were also found to differ in fluidity. The mechanism for the maintenance of such distinct lipid domains in the plasma membrane is not well understood but some intramembranous and submembranous constraints are implicated (Friend, 1982).

5.5. ENDOTHELIUM OF CAPILLARIES

Simionescu et al. (1981a) have demonstrated the existence of micro-domains on the plasma membrane of fenestrated capillaries in pancreas and jejunum, that is, on the blood front. These microdomains represent unequal distribution of anionic sites localized by labeling with cationinized ferritin (MW 480,000; molecular diameter 11 nm) and a smaller cationic probe (alcian blue; MW 1,300; molecular diameter 2 nm). Perfusion in situ with a number of enzymes (hydrolases) indicates that these anionic sites on the blood front are rich in sulfated glycosaminoglycans, most probably heparan sulfate or heparin, whereas the rest of the endothelial surface is of mixed nature (Simionescu et al., 1981a, b). It is likely that such a functional compartmentalization is a common feature of the vascular endothelium.

5.6. PRESYNAPTIC AND POSTSYNAPTIC PLASMA MEMBRANES

Morel et al. (1982) have reported that the presynaptic plasma membrane fraction of the cholinergic synapse from *Torpedo* contains a major protein of MW 67,000. Synaptic vesicles and the postsynaptic membrane did not show the presence of the protein. The pre- and postsynaptic membranes could be further characterized by the pres-

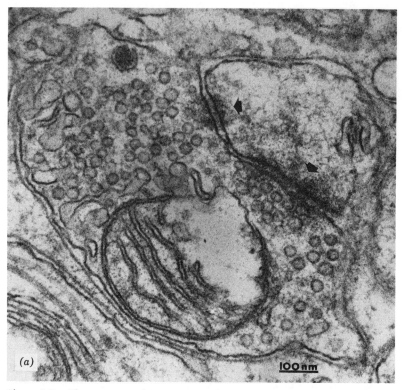

Figure 5.3a The structure of a synapse in the mouse cerebral cortex prepared by fixation in glutaraldehyde and OsO_4. The thin sections were routinely stained with uranyl acetate and lead citrate. Arrows are in the postsynaptic structure and indicate concentration of densely staining material associated with the synaptic membranes. An example of densely staining material in association with the postsynaptic membrane only is shown in Fig. 5.3b.

ence of Ach receptors in the postsynaptic and few IMPs on the fractured face of the presynaptic membranes.

In the brain also, presynaptic and postsynaptic membranes can be distinguished on the basis of the distribution of IMPs (Malhotra et al., 1975). The postsynaptic membrane also stains intensely with tannic acid (Figs. 5.3a, 5.3b, and 5.4). The antibody to calmodulin shows a concentration of this protein underlying the postsynaptic membrane (Cheung, 1982a); it has a characteristic glycoprotein, antigen PSD95 (Nieto-Sampedro et al., 1982).

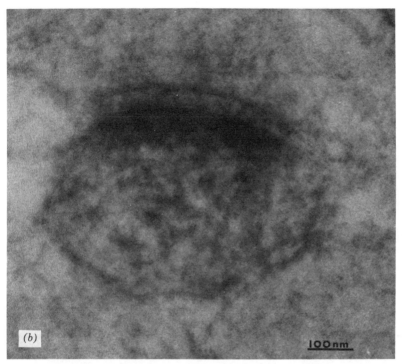

Figure 5.3b Electron micrograph of a synapse in the mouse cerebral cortex stained with tannic acid during fixation with glutaraldehyde. The tissue was subsequently fixed in OsO_4. The sections were examined without further staining in heavy metal salt solutions. The heavy staining of the postsynaptic membrane is reminiscent of labeling with antibodies for calcium binding protein, calmodulin (Cheung, 1982a, b) and calmodulin binding protein (Wood et al., 1980).

The presynaptic membrane is further specialized to facilitate release of neurotransmitter by exocytosis of synaptic vesicles. There are ~700 such sites on the membrane of the axon in a frog neuromuscular junction. The transmitter release can be blocked by botulinum toxin (BOTX). The action of the toxin appears to be indirect, possibly through the release of some intracellular products that interfere with the transmitter release sites, rather than directly blocking the release sites in an "all or none" fashion. These remarks are based on intracellular electrical recordings from frog muscle fibers during an advanced stage of BOTX paralysis. Even though the quan-

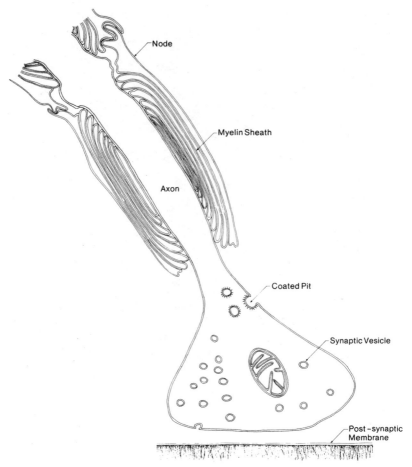

Node

Myelin Sheath

Axon

Coated Pit

Synaptic Vesicle

Post –synaptic
Membrane

Figure 5.4   A highly diagrammatic representation of the terminal part of an axon and an asymmetric postsynaptic membrane. The latter is shown with a layer of uniform thickness of fibrous material on the cytoplasmic surface, but the distribution of this material could be extremely variable. (Not to scale)

tal response had been reduced by 4–5 orders of magnitude below normal, the Ach release could be recorded (e.p.ps, end plate potentials, Gundersen et al., 1981). It should be pointed out that while it is generally believed that Ach is released at the synapse by exocytosis of presynaptic vesicles, the possibility has been considered that a cytoplasmic Ach is also released at the synapse. By using a sensitive chemiluminescent technique, Isräel et al. (1981) measured Ach release from synaptosomes from the electric organ of the *Torpedo* after freezing and thawing. (The vesicular Ach resists treatment to be liberated.) Osmotically shocked synaptosomes could be filled with Ach of a known concentration; most of this Ach was released upon stimulation of the synaptosomes by a calcium ionophore.

There are $\sim 1500 \pm 300$ intramembranous particles per μm^2 at the presynaptic zone (active zone) where exocytosis takes place as compared to 960 ± 450 particles per μm^2 outside the active zone in the giant synapse of the squid *Loligo pealei*. Also, the size of the particles differs in these two locations, that is, 9.2 ± 2 nm versus 8.0 ± 2.5 nm. The active zone particles are thought to represent calcium channels since this zone is specialized for exocytosis (Pumplin et al., 1981).

5.7. NODE OF *RANVIER* AND ION CHANNELS

The axolemma at the node of *Ranvier* in myelinated nerves provides a well-defined example of functional domains in the plasma membrane, as sodium channels are clustered in the nodal region whereas the potassium channels are localized in the axolemma throughout the entire internode, there being no sodium channels in the internodal axolemma (Brismar, 1982; Waxman and Foster, 1980b, Fig. 5.4). By using a specific ligand, ^{3}H-saxitoxin, Ritchie and Rogart (1977) estimated a density of $\sim 12,000/\mu m^2$ for sodium channels at the nodes as compared to less than $25/\mu m^2$ at the internode in rabbit myelinated axon. The detection of potassium channels in the internode and paranode axolemma was made feasible by demyelinating the fibers with 0.1–0.2% lysolecithin (Chiu and Ritchie, 1981). These channels showed characteristic sensitivity for TEA. The nodal freeze-fractured axolemma (E face) shows larger (20 nm) and more (1200–$1600/\mu m^2$) intramembranous particles than the corresponding internodal face (13 nm and 100–$200/\mu m^2$; Kristol et al., 1978). The possibility arises that the nodal intramembranous particles contain sodium channels.

In myelinated axons, the passage of nerve impulses takes place in a saltory manner, and the nodes are the sites of large inward sodium currents (Huxley and Stampfli, 1949). Estimates of single channel conductance give a value of 108 mA/cm^2 for peak sodium current density (membrane conductance 1500–2000 mS/cm^2) for frog node as compared to 2 mA/cm^2 (membrane conductance 55–70 mS/cm^2) for frog skeletal muscle (reviewed by Ritchie and Rogart, 1977). However, the function of the potassium channels in the internode of the myelinated axon is not known as they are superfluous for impulse conduction in the myelinated axon (Brismar, 1982).

The differentiation of axolemma into nodal and internodal regions appears to precede the formation of the myelin sheath; regions of high concentration of sodium channels, followed by cytochemical staining with ferric ion-ferrocyanide (Waxman and Foster, 1980a) or ^{125}I-labeled neurotoxin II from scorpion (Berwald-Netter et al., 1981), appear before the Schwann cells and begin to form the myelin sheath. Such a finding is of obvious value in understanding the control and mechanism of the development of functional domains in membranes.

The existence of nonuniformly distributed sodium channels was demonstrated in the cultured spinal cord neurons by autoradiographic analysis of ^{125}I-scorpion toxin binding sites, there being approximately seven-fold more toxin binding sites on some neurites than on the cell body. Such a distribution of sodium channels is thought to be related to the lower threshold for generation of action potential at the axon rather than the cell body. In contrast to the spinal cord neurons, the mouse neuroblastoma cell lines N18 and C1300 showed a uniform density of ^{125}I-scorpion toxin binding sites over cell body and neurites.

5.8. PLASMA MEMBRANE OF SPORANGIOPHORE OF *PHYCOMYCES*

The sporangiophores of the fungus *Phycomyces blakesleeanus* are giant unicellular cylindrical aerial hyphae that grow from mycelia when dormant spores are activated to germinate (Bergman et al., 1969). The rate and direction of growth are governed by the intensity and orientation of the incident illumination. The growth takes place in the growing zone which extends ~3 mm below the sporangium (Fig. 5.5); the growing zone also contains an as yet unidentified photoreceptor (see Malhotra and Tewari, 1982).

The plasma membrane of the growing zone of the sporangiophore differs from the plasma membrane of the nongrowing zone in the number and distribution of IMPs. The number of these IMPs decreased when the sporangiophores were grown in the dark (Tu and Malhotra, 1975). The relevance of this change to the mechanism of photoreception is not known since neither the nature of the photopigment nor its location are certain.

Figure 5.5 A photograph of *Phycomyces* sporangiophore which is a large unicellular structure. The growing zone extends 2–3 mm below the sporangium, which appears as a rounded structure at the end of the cylindrical sporangiophore (Bergman et al., 1969).

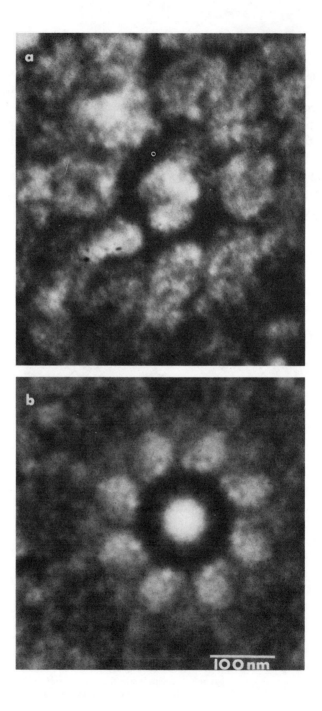

The plasma membrane of the growing zone also contains adenylate cyclase, which was found to be lacking histochemically from the nongrowing zone (Cohen, personal communication, 1982). The growing zone also contains membrane-bound chitin synthetase, which is involved in the biosynthesis of chitin, a major constituent of the cell wall of this fungus.

Based on the observations made on germinating spores cultured in the presence of cAMP in the medium, the distribution of IMPs in the plasma membrane is altered (Tu and Malhotra, 1977). (Besides thickness of the cell wall, intracellular glycogen contents and calcium distribution also undergo changes in the presence of cAMP.) From the above, it is reasonable to conclude that the plasma membrane of the sporangiophore of this fungus consists of functional domains in an otherwise continuous membrane. The molecular structure of these domains remains to be investigated.

The plasma membrane of the germinating spores also has characteristic IMPs that have nine subunits (Fig. 5.6), but their function is not known. It should be of interest to use *Phycomyces* as a model system for studies on photoresponse, germination, and growth of membranes.

Figure 5.6 Showing nine subunit structure of an intramembranous particle (30–35 nm) on the E face of the plasma membrane of dormant spore of the fungus *Phycomyces*, in a freeze-fractured replica of a rotary-shadowed preparation (a) subjected to photographic rotational filtering (b) (Markham et al., 1963). There are eight subunits arranged in a ring around a central subunit. These large IMPs disappear when the spores germinate; their function is not known. Reproduced with permission from Malhotra and Tewari (1982).

6

Intercellular Communication (Cell-to-Cell Junctions)

6.1. INTRODUCTION

One of the best morphologically discernible aspects of the functional domains in organized animal tissues is the establishment of cell-to-cell junctions. Their structure and function are of current interest because at least some of them are involved in regulating developmental events and establishing morphological and physiological contacts between adjoining cells of a tissue.

6.2. INTERCELLULAR JUNCTIONS

In organized tissues, contacts between cells are likely to play vital roles in growth, differentiation, and coordination of cellular activi-

43

ties, in addition to their mechanical role in keeping cells together (Geiger, 1981). With the developments in electron microscopy, our understanding of the structure of cell junctions has progressed so as to facilitate their recognition, but their composition and molecular architecture are not yet known. The best-known cell junctions are the gap junctions which are formed by the close approximation of the plasma membranes of apposed cells and are implicated in inter-cellular coupling, that is, in the existence of low-resistance passages for ions and small molecules between cells. In view of their postu-lated role in the control of development, coordination of cellular functions, and electrical transmission, the gap junctions are de-scribed in some detail in this section (Section 6.3).

Apart from the gap junctions, the other types of cell junctions most readily recognized in epithelia are: (1) tight junction (zonula occuldens), (2) intermediate junction (zonula adhaerens), (3) desmo-somes (macula adhaerens), and (4) septate junctions. The first three named junctions in different epithelia of the rat and guinea pig have been described in detail (Farquhar and Palade, 1963). The septate junctions were first described in invertebrates (Gilula et al., 1970; Wood, 1977) but have also been reported from vertebrates (Friend and Gilula, 1972; Lasansky, 1969). The intermediate junction is characterized by the presence of ~20 nm-wide intercellular space and follows the tight junction. Together with the desmosomes, the intermediate junctions are thought to serve in intercellular attach-ment (Farquhar and Palade, 1963; Fig. 5.1).

The name *tight junction* implies that the junction may form an impermeable seal, particularly in organs that maintain chemical po-tential gradients between the lumen and subepithelial spaces, for example, the kidney and urinary bladder (see Farquhar and Palade, 1963). However, recent physiological studies using tracers, such as mannitol, and on route for water transport in different tissues, have led to the distinct possibility that the tight junctions may not really be tight but leaky junctions. The historical background and experi-mental evidence for water transport is discussed in a review by Oschman (1980). Tight junctions are formed by the lateral close apposition (fusion) of the plasma membranes of the adjacent cells to the extent that the intervening intercellular space is obliterated. They can be readily recognized in the electron micrographs of

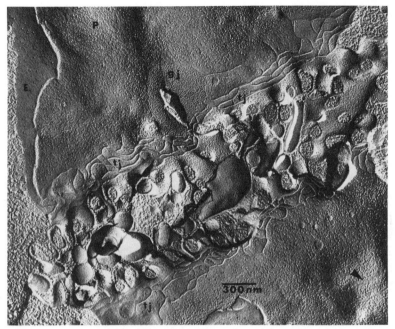

Figure 6.1 Freeze-fracture preparation of mouse liver showing gap junction (gj) and tight junction. The gap junctions are composed of aggregation of IMPs, and the tight junctions appear as anastomosing ridges and furrows.

freeze-fracture replicas, where they appear as anastomosing ridges and grooves on the complementary fractured faces of the plasma membrane (Fig. 6.1). The ridges appear to be formed by alignment of particles. Attempts have been made to correlate the structure of the tight junctions with their permeability properties without a general agreement so far (see Møllgard et al., 1979). However, it appears that discontinuities in the ridges may be related to the leakiness of the epithelium. For example, in the choroid plexus, which functions in the secretion of cerebrospinal fluid, the tight junction of the epithelium shows discontinuities in the ridges. These discontinuities are interpreted as pores in the junctional strands. The choroid plexus junction is regarded as leaky, as compared to the tight junctions in the small intestines. The latter show no discontinuities in the ridges in the freeze-fracture replicas (Van Deurs and Koehler, 1979). The possibility has been mentioned that in invertebrates septate junctions may also serve the role that tight junctions and intermediate junctions perform in vertebrates (Wood, 1977).

Desmosomes that are regarded as the intercellular attachment sites (Farquhar and Palade, 1963) have been recently isolated and subfractionated. Whole desmosomes were purified from the bovine muzzle epidermis and when run on SDS-polyacrylamide gel electrophoresis (SDS-PAGE), showed 12 major bands. Such desmosomes were subfractionated through centrifugation on metrizamide gradients to yield "cores" containing intercellular material. The intercellular material contains almost all of the glycoproteins of the desmosome. The nonglycoprotein material stripped from the cores during centrifugation contains fibrillar plaques. SDS-PAGE revealed four proteins of MW 230,000, 205,000, 86,000, and 82,000 and some prekaratins. Monoclonal antibodies revealed that the proteins of MW 230,000 and 205,000 had pronounced antigenic similarity as did those of 86,000 and 82,000. These antibodies preferentially stained the cell borders (in immunofluorescence microscopy) in the frozen sections of the epidermis, thereby indicating that these proteins are specific to the intracellular desmosomal plaques and that the filaments may serve to link the cytoskeletal tonofilaments with the plasma membrane (Gorbsky et al., 1981). The findings that the intercellular core of the desmosomes contains three distinct glycoproteins (MW 150,000, 115,000, and 105,000) suggest that they may serve as the intracellular adhesive (Gorbsky and Steinberg, 1981).

Besides glycoproteins, which are suspected to play a role in cell-to-cell adhesion, it is likely that the four types of cell junctions (gap junctions, tight junctions, desmosomes, and intermediate junctions) may contribute variously by an as yet unspecified mechanism to cell-to-cell adhesion (Garrod and Nicol, 1981). It is also likely that initial events in adhesion between cells are facilitated by van der Vaals forces and electrostatic forces (Garrod and Nicol, 1981).

6.3. GAP JUNCTIONS

The physiological basis of the existence of gap junctions emerged from studies of the mechanism. of nerve transmission as Furshpan and Potter (1957) described electrical transmission (ionic coupling or low-resistance pathways, intercellular communication) in the nervous tissue of crayfish. The morphological correlates of electrical coupling were reported by Bennett et al. (1963) and Robertson (1963). This aspect of cell biology gained prominence when Loewenstein (1981) reported the presence of low-resistance pathways connecting apposed cells in nonexcitable organized epithelial tissues, a phenomenon which has been demonstrated in many developing and adult animal tissues, both excitable and nonexcitable (Bennett and Goodenough, 1978). Low-resistance passages between apposed cells have been found nearly everywhere in organized animal tissues, even in primitive multicellular organisms (Loewenstein, 1981). It is now commonly believed that gap junctions are the morphological structures that contain hydrophilic channels across the plasma membranes connecting the cytoplasm of apposed cells (Loewenstein, 1981). The evidence for the physiological role of gap junctions is derived from correlative studies involving electrophysiological studies (ionic coupling), passage of fluorescent dyes across plasma membranes without leakage into the extracellular material (dye coupling), and passage of metabolically important molecules (metabolic coupling).

To elucidate the role of gap junctions, Gilula et al. (1972) demonstrated that the metabolic coupling (and ionic coupling) in cultures of Chinese hamster cell lines (fibroblasts) is associated with the formation of gap junctions. They used three cell lines: (1) Don (which can incorporate from the culture medium purine ^{3}H-hypoxanthine into its nucleic acids, (2) DA, and (3) A9 cells (which lack the enzymatic activity required for such an incorporation of the exogenous purine). Prelabeled Don cells (with carmine or carbon or ^{3}H-thymidine) were cocultured with DA or A9 cells in the presence of ^{3}H-hypoxanthine. Microscopic examination revealed that only when DA cells were in contact with Don cells did DA cells show isotope incorporation (metabolic coupling), though less than the Don cells. Such groups of cells manifested electrical (ionic) coupling and the presence of gap junctions. A9 cells, even when in close contact with the Don cells, did not show metabolic coupling, ionic coupling, or the formation of gap junctions. Apparently A9 cells are deficient and unable to make functional contacts even when in contact with Don cells. However, these experiments clearly reveal the association between the functional contacts and gap junctions in these cultured fibroblasts.

Gutnick and Prince (1981) have demonstrated dye coupling and electrotonic coupling in the neocortical slices of guinea pig. Iontophoretic injection of a fluorescent dye (Lucifer Yellow CH) into single neurons showed that the dye spread to a group of neurons (dye coupling). Intracellular recordings also provided evidence in favor of the existence of electrotonic coupling in neuronal aggregates. In control experiments the dye was ejected from the microelectrode into the extracellular space, and there was no spread into the neurons.

Based upon electrophysiological data on Novikoff hepatoma cells in cultures, the specific junctional conductance was estimated to be 0.78×10^2 mho/cm^2. Such a value is consistent with the existence of hydrophilic channels, $\sim$2 nm in diameter, traversing the junctional membranes. The resistivity of such a channel is comparable to that of the cytoplasm (Sheridan et al., 1978).

Though the gap junctions are discernible in tangential sections of lanthanum-soaked tissues as hexagonally packed arrays (Revel and Karnovsky, 1967), they are most readily recognized in freeze-fracture replicas as an aggregate of intramembranous particles on the P face (E face in some invertebrates; Bennett and Goodenough, 1978) and corresponding pits on the complementary fractured face (Fig. 6.2). The size of the particles varies from approximately 8.5 nm in mouse hepatocyte (Sikerwar et al., 1981) to 15 nm in some inverte-·brates (Berdan and Caveney, 1981). (Molluscan gap junctions essentially resemble the mammalian gap junctions in their structure [Gilula and Satir, 1971].) These particles span the plasma membrane, and particles from the apposed membranes, termed "connexon", are in register and linked to each other (see Caspar et al., 1977). Based upon electron microscopy and low-angle X-ray diffraction studies (Caspar et al., 1977; Makowski et al., 1977), it is thought that the particles have six subunits arranged in an annulus delineating a hydrophilic channel mentioned above. Freeze-etching experiments on isolated lens gap junctions suggest that the particles may not protrude into the cytoplasm (Peracchia and Peracchia, 1980a). Simi-

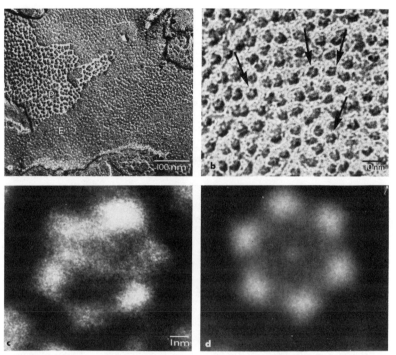

Figure 6.2 Structure of gap junction: (a) Rotary-shadowed replica of mouse liver freeze-fractured tissue with IMPs on the P face and depressions on the complementary E face; (b) a magnified view of P face showing subunit structure of connexons (arrows); (c) a magnified view of a connexon in reverse contrast that has been subjected to photographic rotation enhancement of contrast (Markham's rotation) and shows six subunits (d). Reproduced from Sikerwar et al. (1981).

lar observations have been reported by Hirokawa and Heuser (1981) in the liver gap junctions.

Recent studies on negatively stained rat hepatocyte gap junctions (Fig. 6.3) subjected to three-dimensional Fourier reconstruction (of low-dose electron micrographs) indicate that the six protein subunits are arranged in a cylinder slightly tilted tangentially, enclosing a channel 2 nm wide at the extracellular region. The dimensions of the channel within the membrane were narrower but could not be resolved (Unwin and Zampighi, 1980; Fig. 6.4). A small radial movement of the subunits at the cytoplasmic ends could reduce the sub-

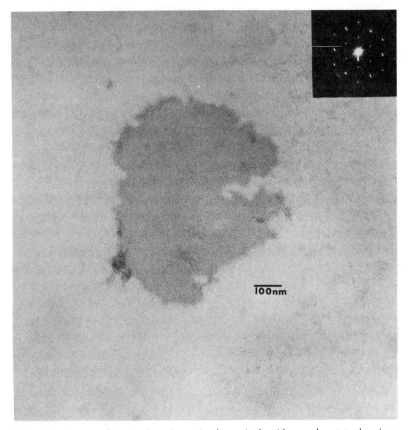

Figure 6.3 A mouse liver gap junction stained negatively with uranyl acetate showing a well-ordered arrangement of the connexons. Inset: optical diffraction pattern showing a hexagonal lattice with a lattice constant of 8.5 nm. (Unpublished observations, Sikerwar and Malhotra)

unit inclination tangential to the six-fold axis and close the channel (Unwin and Zampighi, 1980). Further details of the molecular organization should emerge as preparative methods become available so that high-resolution three-dimensional images comparable to the purple membranes (Henderson and Unwin, 1975) can be obtained.

Recent experiments on rotary shadowed replicas of rapid-frozen gap junctions in situ, combined with the photographic enhancement of contrast by Markham's rotation (Fig. 6.2), confirm a six subunit

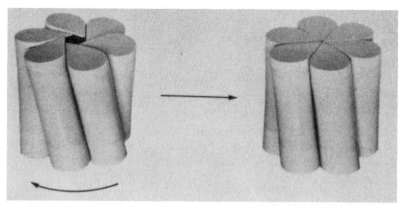

Figure 6.4 Wooden model of the connexon depicting the transition from the "open" to the "closed" configuration. It is proposed that the closure on the cytoplasmic face (uppermost) is achieved by the subunits sliding against each other, decreasing their inclination and hence rotating, in a clockwise sense, at the base. The darker shading on the side of the model indicates the portion that would be embedded in the membrane. The radial displacement of each subunit at the cytoplasmic end would be about 6Å, given the observed inclination change of 5° distributed over its 75Å length. Reproduced with permission from Unwin and Zampighi (1980).

structure in the mouse hepatocyte (Sikerwar et al., 1981). In this tissue as well as in the heart of *Ciona* (Hanna et al., 1981), there are also connexons that have four or five subunits. Whether such a subunit structure exists in vivo is not yet certain, as an apparent tetrameric or a pentameric appearance could arise from an hexameric structure when the latter is viewed at an angle of ~30° (Sikerwar et al., 1981). Such an appearance could also arise in electron micrographs of freeze-fracture replicas since the fracture plane may vary within the bilayer. However, a junction that superficially resembles gap junction and has a tetrameric symmetry is found in mammalian lens fiber cells. This lens junction appears to be a representative of a different class of cell-to-cell junction (Costello et al., 1982) and is not homologous with the gap junctions in most tissues (Hertzberg et al., 1982).

Chemicals of molecular weights of up to 1000 (diameter 1.2 nm), when injected into cells, move across the plasma membrane into adjacent cells (Bennett et al., 1981). The upper limit of the mammalian channel is between 1.6 nm and 2.0 nm, whereas this limit in insects is between 2.0 nm and 3.0 nm. This estimate is based upon studies with fluorescent-labeled oligosaccharides (Loewenstein, 1981). However, even though the cells may be electrically coupled, as for example the mammalian pacemaker cells of the sinoartial node, intracellularly injected fluorescein molecules (MW 320) fail to spread to the adjoining cells. Such an instance raises the possibility that the intercellular channels through which electrical currents (ions) pass may be too small for the passage of small molecules of the size of fluorescein (de Mello, 1980). Also, it is not yet certain whether electrical coupling occurs in the absence of gap junctions, as instances have been cited where the gap junctions have not been detected in electrically coupled tissues. Examples include the longitudinal smooth muscles of the intestine of the dog and of the uterus of the rat (Daniel et al., 1976). Gap junctions may not be readily recognized in freeze-fracture electron micrographs because they may be sparse and made up of a few IMPs.

Notwithstanding the above, it is commonly believed that the gap junctions are the morphological correlates of electrical coupling. Whether these are the only structures that mediate coupling between cells has not been settled. Moreover, the specific role or the nature of the molecules exchanged between coupled cells in vivo is not yet certain. There is evidence from cell culture studies that cAMP is one of the likely candidates in this category. Junctional permeability has been reported to increase with increasing levels of intracellular cAMP concentration (Flagg-Newton et al., 1981).

The evidence for the role of cAMP as a mediator molecule in intercellular communication comes from studies carried out by Lawrence et al. (1978). They cocultured rat ovarian granulosa cells and mouse myocardial cells and investigated intercellular communication between these heterologous cells. Finding that there was electrical coupling between the heterologous cells, they established gap junctions between them. ^{3}H-uridine or ^{3}H-adenine was exchanged between cells when one type of cell was prelabeled, and the isotope was detected in the heterologous cell by autoradiography. Both cell types have characteristic hormonal response to noradrenaline, and the ovarine cells respond to follicle-stimulating hormone (FSH).

Their response is mediated through intracellular cAMP. Thus, in the heterologous cocultures, hormonal stimulation of one cell type produced response in the coupled cell of the other type; that is, stimulation of myocardial cells by noradrenaline produced plasminogen activator, which is the characteristic response of the granulosa cells stimulated by FSH. Similarly, stimulation of the granulosa cells by FSH produced an increase in the beat frequency of the myocardial cells. Since the hormonal response in both these cells is mediated by cAMP, this is the most likely molecule that could be involved in the intercellular communication between the two cell types.

Gap junctions between the oocytes and granulosa cells in mammalian ovary (Anderson and Albertini, 1976) apparently serve to transmit an inhibitor that maintains meiotic arrest in the oocytes. Gap junctions between the follicle cells and the oocytes in *Xenopus* may also serve for the passage of an "initiation factor" from the follicles as the opening and/or assembly of these junctions is effected by gonadotropins (Browne et al., 1979). It is quite conceivable that some of the gap junctions are asymmetrically permeable (Flagg-Newton and Loewenstein, 1980), as are some of the rectified electrical synapses, for example, the synapse between the lateral or medial giant axon and large motor axons in the nerve cord of the crayfish (Furshpan and Potter, 1957, 1959). That is, the transmission occurs unidirectionally from pre- to postsynaptic fiber.

Apart from the precise "information" exchanged between the coupled cells, it is of current interest to understand the physiological control(s) that regulate the permeability of the gap junctions. Ca^{2+} and/or H^+ ions have been implicated in this regulator role (Turin and Warner, 1977). An increase in intracellular concentration of Ca^{2+} has been claimed to uncouple electrically coupled cells of salivary gland of *Chironomus* (Loewenstein, 1981). Bennett et al. (1981) and Spray et al. (1981a) have reported that the conductance of the gap junctions in the embryos of *Ambystoma* and *Fundulus* is dependent on the cytoplasmic pH. The effect of pH was demonstrated when the embryos were exposed to physiological saline equilibrated with 100% CO_2, or external application of the weak acids lactate, propionate, and acetate, whose unionized form diffuses across the plasma membrane to release H^+. No apparent hysteresis in the effect of pH on the junctional conductance was noticed, which thus indicates that an indirect action through, for example, calcium, was unlikely. In support of these findings, Bennett and colleagues found that in the coupled blastomeres the gap junctions were 10 thousand times less sensitive to Ca^{2+} than to H^+. Also, measurement of the intracellular Ca^{2+} using aequorin showed no detectable change in Ca^{2+} during a decrease in junctional conductance induced by acidification. It is likely that in a pathological state in which the plasma membrane of a cell is disrupted, the membrane had uncoupled from its neighboring cells and that this uncoupling may have been produced by Ca^{2+} ions entering from the extracellular medium.

In amphibian embryos, the junctional conductance is sensitive to voltage (Bennett et al., 1981). It is conceivable that voltage-controlled uncoupling might be operational in separating cell populations in amphibian embryos when substantial resting potential differences occur during development (Warner, 1973). However, the voltage-dependence of the junctional conductance does not appear to be a general feature even in various amphibian embryos, whereas the pH dependence is a general property of junctions. Recently, a calmodulin-like protein has been reported to be involved in the uncoupling of cells as an inhibitor of calmodulin, namely, stelazine. It prevented uncoupling in amphibian *Rana* and *Xenopus* embryos, which could be induced by CO_2 after washing out the inhibitor (Peracchia et al., 1981).

The organization of the gap junctions is subject to varying physiological controls. One such control is apparently at the hormonal level. For example, estradiol, enclomiphene citrate (a nonsteroidal estrogen), and FSH produced amplification of the gap junction membrane in the rat granulosa cell layer of ovarian follicles. Ovarian follicles from hypophysectomized rats were subjected to exogenous hormonal stimulation 60 to 90 days after surgery, and the structure of the gap junctions was analyzed 24 hours after the second daily injection (Burghardt, 1981). In *Tenebris molitor,* epidermial cells exposed to 20-hydroxyecdysone showed an elevation of the gap junctional conductance, though no structural alteration in the IMPs could be detected. However, uncoupling of these cells by chloropromazine produced closer packing of the particles as compared to those in the control (Berdan and Caveney, 1981).

It would appear from the above that the gap junctions are dynamic structures whose functional state and formation are dependent upon physiology. Their dynamic nature is best illustrated by the presence of electrical coupling connecting cells not only in the same tissue but also in different embryonic tissues (Sheridan, 1968), whereas in the adult, not all tissues (e.g., skeletal muscle fibers and nerve cells) show electrical coupling. (This subject is further discussed by de Mello, 1980.)

It is also of interest to correlate the structure of the gap junctions with their physiological state. Previous studies carried out in situ or on isolated gap junctions have been interpreted as suggestive of the close packing of the connexons, often in a hexagonal pattern in uncoupled gap junctions as compared to the random or less-ordered packing of the connexons in the coupled state. Such an alteration in the packing of the connexons has been attributed to Ca^{2+} (Peracchia and Peracchia, 1980b). However, such studies have often been carried out on material fixed in glutaraldehyde and cryoprotected by glycerol. Rapid freezing of gap junctions in situ without cryoprotection but with or without prior fixation in glutaraldehyde for up to 1½ hours produced only a slight shrinkage of the interconnexon spacing of the randomly dispersed connexon in the mouse hepatocyte (Sikerwar and Malhotra, 1981). Also, in the gap junctions in the heart of *Ciona* the connexons have been reported to be randomly dispersed when rapidly frozen in a low-resistance state or in a high-resistance state immediately produced by lowering intracellular pH with CO_2 (Hanna et al., 1981). However, when the junctions had been kept in a high-resistance state for 1½ hours, the connexons were hexagonally packed (Hanna et al., 1981). Clearly, further investigations are desirable to understand the structure and composition of the junctions and the organizational changes that may occur in their functioning.

The composition of the gap junctions is under investigation. Apart from the 26,000 MW protein (connexin), Henderson et al. (1979) found that the mouse hepatic gap junctions also contain a large amount of cholesterol that could not be extracted without disrupting the morphology of the junction. The molar ratio of cholesterol and phospholipid in the gap junction was 1.5 ± 0.09 as compared to 0.9 ± 0.1 for the plasma membrane. It was conjectured that cholesterol may contribute to the stability of the gap junction structure through interactions with the protein.

As mentioned above, it is currently believed that gap junctions from different sources contain a predominant type of polypeptide, between 25,000 and 30,000 molecular weights (see Bennett et al., 1981; Henderson et al., 1979). Beyond that, the composition of the gap junctions is not established. For example, gap junctions of rabbit hearts show five peptides on SDS-PAGE; these are of 34,000, 33,000, 31,500, 30,500, and 29,000 MW (Manjunath et al., 1981). Some of the differences reported in the polypeptide composition of the gap junctions from various tissues may result from varying proteolysis, but it is also conceivable that such differences reflect biochemical differences in the junctions of different tissues. Such an assumption is supported by recent findings that eye lens gap junctions are affected differently by Ca^{2+} than are liver gap junctions. The intramembranous particles on the etched face of the gap junctions of the lens exposed by incubation in hypertonic solution and/or Ca^{2+} free solution do not crystallize upon exposure to calcium, whereas liver gap junctions do crystallize under similar conditions (Hirokawa and Heuser, 1981). Also, fixation by glutaraldehyde seems to affect differently the gap junctions in these tissues (Peracchia and Peracchia, 1980b; Sikerwar and Malhotra, 1981). This is further borne out by the observations that the gap junctions in lens and liver have no homology in their N-terminal amino acid sequences (Nicholson et al., 1980).

Hertzberg et al. (1982) have reported that the rat liver gap junctions and the morphologically similar junctions in bovine lens fiber cells are distinct in their polypeptides: The liver junction has a major polypeptide of 27,000 MW, whereas the lens fiber junction has its major polypeptide of 25,000 MW. These polypeptides are not homologous when compared by partial peptide mapping in SDS: There is no antigenic similarity between the two polypeptides. It would therefore appear that the junctions between the lens fiber cells are not related to the gap junctions in most other cells. Such an assumption is supported by the findings that the polypeptides from the gap junctions of mouse liver, rat liver, bovine liver, and rat ovary are homologous (Hertzberg and Gilula, unpublished observations; quoted by Hertzberg et al., 1982).

Gap junctions apparently serve in the development and growth of organized tissues, and regulate their cellular functions, but the mechanism(s) involved remain to be understood. Gap junctions in developing insects present a well-suited system for probing their role in development (Warner and Lawrence, 1982; Weir and Lo, 1982). The imaginal disks, for example, are groups of epithelial cells in the larvae from which cuticular structures arise in the adult. These disks are compartmentalized in respect of their developmental fate. For example, the wing imaginal disks are compartmentalized between the anterior and posterior portions, and the gap junctions between these two compartments were found not to be freely permeable to Lucifer Yellow CH (MW 450), whereas the gap junctions between cells within one developmental compartment were freely permeable to the fluorescent dye.

6.4. PORE-FORMING PROTEIN IN BACTERIA

A bacterial protein somewhat similar in function to the connexin of the gap junctions exists in the outer membrane of Gram-negative bacteria. This protein, referred to as porin (matrix protein, MW 36,500), forms transmembrane hydrophilic channels that allow diffusion of solutes of MW 700 (see Schindler and Rosenbusch, 1982). The channel exists in either a "closed" or "open" state, and in the functional state it forms trimers with the possibility of three pores per trimer. The trimer formation also requires the presence of bacterial glycolipid (Schindler and Rosenbusch, 1981). When detailed analysis becomes available, this transmembrane protein may turn out to have structural features comparable to that of the connexin. No obvious hydrophobic domains in the amino acid sequence of porin have been detected. It has a predominantly β-configuration, it is highly charged, and it is apparently almost entirely embedded in the hydrophobic environment of the membrane with a single amino group available for reaction with pore impermeant reagents (losin-isothiocyanate, MW 705; Schindler and Rosenbusch, 1982).

It is of interest that this protein has been crystallized by using the detergent octylglucoside; a structural analysis is awaited (Garavito and Rosenbusch, 1980). (The other membrane proteins that have been crystallized so far are the bacteriorhodopsin [Chapter 9] and cytochrome-c-oxidase cytochrome c complex [reviewed by Henderson, 1980].)

6.5. INTERCELLULAR COMMUNICATION THROUGH EXTRACELLULAR SPACE

Intercellular communication in organized tissues could take place by the transmission of signals through extracellular space. For instance, it has been reported in leach and snail that the K released by the neurons during nerve activity stimulates glycogen synthesis in the surrounding glial cells (Pentreath and Kai-Kai, 1982). Such a role for extracellular K has been demonstrated by stimulating nerves with electrodes or by the addition of K to the saline solution in the presence of 2-deoxy-D[1-3H] glucose (2-DG) in isolated preparations of ganglia. 2-DG is selectively metabolized into glycogen, and the glial cells showed an increase in the incorporation of 2-DG into glycogen of up to four times with a maximal increase at a concentration of K at ~4 mM above normal. (4 mM concentration is within the normal range of extracellular K during nerve activity [Kuffler and Nicholls, 1976].) Such an interaction between neurons and glial cells has an important implication because in the central nervous system where the blood–brain barrier may prevent hormonal regulation of glycogen metabolism, local controls within the nervous system could ensure continued energy supply (Pentreath and Kai-Kai, 1982).

   Another well-known example of intercellular communication through extracellular space is the inhibition of the Mauthner cell (Furukawa and Furshpan, 1963). These neurons in the goldfish brain manifest hyperpolarization that is dependent upon extracellular current generated by cellular elements in the vicinity of the Mauthner cell.

7

Nicotinic Acetylcholine Receptor (Ach Receptor)

7.1. INTRODUCTION

Nicotinic Ach receptors, henceforth referred to as the Ach receptors, have been most well known from the postsynaptic membrane of the vertebrate neuromuscular junction and from the homologous electroplaques in the electric organ of *Torpedo* and *Electrophorous*. These receptors can be isolated in quantity from the electroplaques and hence are suitable for biochemical and biophysical analyses, which form the basis of our current understanding of the structural features. They are also abundant in the myotubes which can be cultured and examined by labeling with fluorescent markers. Furthermore, the adult skeletal muscle, when denervated, is well known to manifest supersensitivity to Ach (see Axelsson and Thesleff, 1959; reviewed in Malhotra, 1981; Miledi, 1960), and this sensitivity results from the muscle's incorporation of newly synthesized Ach receptors into the sarcolemma outside the postsynaptic region (Fambrough, 1979). The extrajunctional Ach receptors are very similar to the junctional Ach receptors (see below). In the embryonic muscle, as well as in the denervated muscle, the Ach receptors, which may be clustered in patches, form *hot spots,* regions of the sarcolemma that manifest an Ach response several degrees higher than elsewhere on the sarcolemma (Fishbach and Cohen, 1973). The hot spots may vary from less than 1 μm to 30 μm and have served as useful experimental material for the morphological correlation of Ach receptors (Malhotra, 1981; Tipnis and Malhotra, 1980).

Ach receptors are also of interest for immunological studies because in the autoimmune disease *Myasthenia gravis* in humans, the Ach receptors are degraded by antibodies against Ach receptors (see Drachman et al., 1979; Fuchs, 1979). It is of obvious interest to develop a means of suppressing the disease.

Lithium, which is also used in the treatment of manic patients (Tosteson, 1981), has been reported to reduce the number of Ach receptors in skeletal muscle, an effect that may be related to its properties as a cation (Pestronk and Drachman, 1980b). Certain antidepressant drugs, which primarily act on the central nervous system, also affect the conduction of ions through Ach receptors (Schofield et al., 1981). An understanding of the action of such drugs should facilitate advances in the functioning of these receptors.

Apart from the skeletal muscle and the electric organ of the fish, Ach receptors have been described from most parts of the brain, in ciliary ganglion of the chick, and in sympathetic neurons (reviewed by Morley and Kemp, 1981). But it is not yet certain how far these receptors differ from those in the skeletal muscle. It appears that they may differ a great deal (Section 7.8). Also, Rehm and Betz (1981) reported that phospholipase c treatment of retinal membranes from newly hatched chickens resulted in a twofold increase in ^{125}I-α-BGT binding to the membranes. These unmasked α-BGT binding sites were indistinguishable in their pharmacological characteristics from the control membranes. However, phospholipase-c-treated membranes showed that one-half of the toxin binding sites were inhibited by CO^{2+} (greater than 1 mM), whereas the controls were insensitive to up to 8 mM CO^{2+}. The origin of the cryptic α-BGT binding sites is not understood. However, it is conceivable that the cryptic and the accessible α-BGT binding sites are located on the same Ach receptor molecule, as there are two known agonist binding sites in muscle and fish electric organs. Embryonic chick pectoralis muscle showed only a moderate increase (~30%) in α-BGT binding sites after treatment with phospholipase c. These differences in muscle and retina may simply reflect differences in the fractional occupancy of their α-BGT binding sites by the toxin.

Ach receptors have also been found on human lymphocytes. Whether these receptors are abnormal in myasthenic patients or are in any way related to this autoimmune disease remains to be learned (Richman and Arnason, 1979).

Our studies on Ach receptors have been greatly facilitated by the availability of postsynaptic neurotoxins from snake venom. One such toxin is the well-known α-Bungarotoxin (α-BGT) which binds specifically and irreversibly to the nicotinic Ach receptors in muscle and electroplaques (Chang and Lee, 1963; Lee, 1979). The binding of the neurotoxin to the receptor is thought to be by multiple electrostatic and hydrophobic interactions (Lee, 1979). Also, α-BGT has been conjugated to tritium, ^{125}Iodine, or horseradish peroxidase or FITC for characterization of the receptors. Immunochemical techniques are also being applied for structural studies (Klymkowsky and Stroud, 1979) as well as for cross-reactivity between possible subunits of Ach receptors (Lindstrom, 1979). A number of agonists and antagonists and affinity agents for Ach receptors are available and in wide use.

The term *Ach receptor* is applied to the entire macromolecular membrane complex, which includes the Ach receptor protein and the acetylcholine ionophore, described as the acetylcholine regulator (Heidmann and Changeux, 1978). The binding of the Ach to its recognition site is followed by a conformational change of the Ach receptor that leads to the transitory opening of an ionic channel through the membrane (Heidmann and Changeux, 1978; Karlin et al., 1979; Nachmansohn, 1955). The ionic channel is nonselective; that is, Na^+, K^+, and Ca^{++} move through it. The regulation of the opening and closing of this channel is of interest to electrophysiologists. With the development of "patch formation techniques" single ion channels can be experimentaly manipulated (Neher and Sakmann, 1976; reviewed by Lester, 1981); one can expect these techniques to yield insights into the mechanism of opening and closing of these channels. The conductance of a single open channel has been measured at 25 pS and the pore open-time averages around 1 ms: the current flowing through the receptor protein is ~2.5 pA (see Stevens, 1980). Hamill and Sakmann (1981) made use of cell-free membrane patches from uninnervated embryonic muscle of the rat and showed that the Ach receptor channels manifest various independent states of conductance: one main conductance state; another, a "substrate" of lower conductance; and the third, the closed state. It has been suggested that the subunits comprising the receptor complex could rearrange themselves to form the different open states of a channel. Furthermore, Colquhoun and Sakmann (1981) reported that by using the acetylcholine-like agonist suberyldicholine, the Ach channel opens and closes several times during a single agonist receptor occupancy. On the average, a single open burst was interrupted 3.1 times by gaps lasting 45–70 μs. These experiments were done on the cutaneus pectoris muscle of the frog *Rana temporaria*.

The number of Ach receptors has been determined by using estimated conductance/μm^2, electron microscopy of negatively stained receptor-rich electroplaques and freeze-fracture, and autoradiography of ^{125}I-α-BGT-labeled electroplaques and skeletal muscle. There is a wide variation in the estimates reported from different sources (see reviews by Heidmann and Changeux, 1978; Tipnis and Malhotra, 1981): in electroplaques of *Electrophorus electricus* there are some 50,000 ± 16,000 receptors/μm^2. In mammalian muscle the estimates vary from 8700/μm^2 to 46,000 ± 27,000/μm^2, the highest density being at the top of the junctional folds. The extrajunctional receptor density increases from 5 receptors/μm^2 to ~1700 receptors/μm^2 in denervated diaphragm (Fig. 7.1).

The Ach receptor glycoprotein is an integral membrane protein that is synthesized in association with membrane-bound polyribosomes and passes through the Golgi complex before it is incorporated into the sarcolemma (Fambrough, 1979). The newly synthesized Ach receptor subunits appear to require approximately 15–30 min-

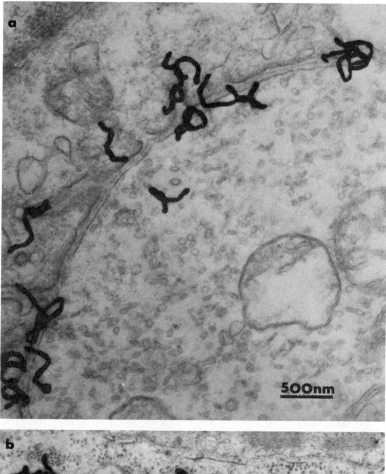

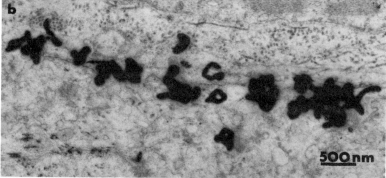

Figure 7.1 Autoradiographs of the rat EDL labeled with ^{125}I-α-BGT showing concentration of silver grains on the sarcolemma in the region of the neuromuscular junction in normal muscle (a) and in the nonjunctional region in the denervated muscle (b). Reproduced with permission from Tipnis and Malhotra (1981).

utes before they acquire α-BGT binding activity. This finding implies that a pool of inactive precursors exists for some time (Merlie and Sebbane, 1981). Futhermore, by using a cell-free system suitable for the synthesis of Ach receptor proteins, Merlie et al. (1981) demonstrated that one of the subunits (α subunit; see below) of the receptor is synthesized on the membrane-bound polyribosomes. Studies are in progress to study the mechanism(s) involved in the synthesis of the Ach receptor.

The Ach receptor belongs to that category of polypeptide receptors on the plasma membrane that does not necessarily require internalization of the ligand or the receptor to effect cellular response (Kaplan, 1981). Other examples of similar classes of receptors are receptors for hormones (e.g., epinephrin, glucagon) and humoral nonhormonal agents (e.g., lgE, Met-Leu-Phe peptide). In contrast, the receptor for low-density lipoprotein (LDL) is internalized after ligand interaction to cause a change in cellular response. If ligand-receptor internalization does not occur, the result is abnormal cell behavior (as in hypercholesterolemic individuals, Chapter 6). The receptor in this second category is likely to be recycled rapidly after delivering the ligand to the cell. In the first-mentioned category, exemplified by the Ach receptor, the receptor may be used once (see Kaplan, 1981), but could be recycled as a part of membrane recycling. However, in the case of the Ach receptor, receptor degradation seems to take place, as evidenced by the appearance of α-BGT binding sites in lysosomes (Fambrough, 1979; Tipnis and Malhotra, 1981).

Recycling of the receptors may not necessarily require a delivery of the ligand to the lysosome, as demonstrated for the asialoglyco-protein receptor of the hepatocyte (Baenziger and Fiete, 1982). Under experimental conditions, which include the omission of Na^{2+} but the presence of 0.15 M K^+, the receptor is able to internalize the ligand, but the lysosomal degradation of the ligand does not occur. The ligand was found in uncoated vesicles and estimates indicated that 20 ligand molecules per high-affinity receptor were observed, which suggests that the receptor reutilization took place without the delivery of the ligand to the lysosomes.

7.2. REMARKS ON CHEMICAL COMPOSITION AND SUBUNIT STRUCTURE

One of the characteristic features of the Ach receptor complex in *Torpedo* electroplaques appears to be its rather high cholesterol (47 mol %) and low sphingomyelin (<2%) of the total phospholipid (see Changeux and Dennis, 1982). ESR studies indicate that there is lipid intercalated with the Ach receptor protein, and it is able to move freely within the interstices of the receptor complex. Also, the lipid apparently remains mobile when cholinergic agonists occupy the receptor binding site, whereas local anesthetics enhance the fluidity. It is not clear whether there are major differences in the lipid composition of the Ach receptors from different sources. As Bridgman and Nakajima (1981) reported, in contrast to the above statement on *Torpedo,* the Ach receptor complex is relatively low in cholesterol. This conclusion is based upon freeze-fracture studies of cultured *Xenopus* embryonic muscle cells treated with filipin, digitonin, and saponin. The freeze-fractured plasma membranes of the treated muscle contain 19–40 nm protuberances and dimples that are absent from the region of the intramembranous particles that represent the Ach receptor molecules. The 19–40 nm complexes were thought to contain sterol-specific complexes.

All four subunits of the receptor contain carbohydrates, but their function does not appear to be known.

After several years of controversial reports (Changeux and Dennis, 1982; Conti-Tronconi and Raftery, 1982; see the review by Malhotra, 1981), it is now generally agreed that the electroplaque Ach receptor is a complex of four subunits, MW (α) 40,000, (β) 50,000, (γ) 60,000, (δ) 65,000, in stoichiometry of $2:1:1:1$. This receptor complex has a MW of 255,000–270,000 and an S value of 9. These data are based on studies of the Ach receptor purified from *Torpedo californica* by sucrose density gradient centrifugation and subsequently analyzed by SDS-PAGE (Raftery et al., 1980). The four subunits show a good deal of homology in their amino acid sequence determined for 56 residues; all four subunits have the same amino acid at 11 of the 56 positions. Also, in several other positions, either two or three of the residues are identical. In only 4 positions are the residues in all four subunits distinct (Conti-Tronconi and Raftery, 1982). These four polypeptides may have originated from a single ancestral gene by fourfold gene duplication early in evolution. (In this respect, the Ach receptor is comparable to the subunits of light and heavy chains of the immunoglobulins, which are thought to have descended from a common ancestral gene [see Raftery et al., 1980].) It also seems most likely that the muscle receptors have wide structural homology to the electroplaque receptor, as a monoclonal antibody against the Ach binding site of the receptor from *Torpedo* bound to the Ach receptors in primary muscle cultures of fetal calf (Conti-Tronconi et al., 1981) and chicken, mouse, and rat (Mochly-Rosen and Fuchs, 1981). Also, antisera raised against each of the four subunits of *Torpedo* cross-reacted with Ach receptors from muscle of human, rat, fetal calf, and monkey (Lindstrom et al., 1978). Furthermore, in rat muscle in vivo, the Ach receptor cross-reacting with antisera against the four subunits caused a decrease in Ach receptor contents, which thus indicates that the cross-reacting determinants are exposed on the extracellular surface. Further studies on the subunit structure are likely to shed light on the variations in the Ach receptors in the muscle from different sources. Also, there may be differences in the extent of lipid (see below) and protein beyond the four polypeptides associated with the receptor in muscle of different species. In contrast to the ~8.5 nm intramembranous particles in the electroplaque, the particles that represent the Ach receptors in muscle from frog, chick, and rat are shown in muscle freeze-fracture electron micrographs to reveal differences in size of from 10 to 19 nm (reviewed by Malhotra, 1981). How far these differences arise from differences in preparative techniques is not yet certain. There may also be variations in the aggregation state of the receptor monomers in different muscle.

The junctional and extrajunctional Ach receptors in muscle are thought to be similar molecules. They show subtle differences, some of which could arise from differences in molecular environment rather than in molecules. Commonly cited differences include: (1) an isoelectric focusing peak (pH 5.1 vs pH 5.3; Brockes and Hall, 1975) that suggests a difference of 10–30 charges per receptor molecule

(Fambrough, 1979) and (2) Ach-induced channel lifetime being shorter in innervated ($\sim$1 ms) than in the denervated muscle by four- to fivefold (reviewed in Malhotra, 1981); the half-life of junctional receptors is longer ($\sim$6 days) than that of the extrajunctional receptors ($\sim$18 hours; see Fambrough, 1979). However, differences in antigenic determinants between the two types of receptors are suggested from results obtained by Weinberg and Hall (1979). They reported that the extrajunctional receptors contain determinants that could be detected by sera from myasthenic patients. All of the determinants of the junctional receptors detected by myasthenic sera were present on the extrajunctional receptors so that myasthenic sera contained two classes of antibodies, one directed against determinants present on both junctional as well as nonjunctional receptors and the other directed against determinants present only on the extrajunctional receptors. It is noteworthy that antisera from animals immunized with rat extrajunctional receptors or eel or *Torpedo* Ach receptors could not detect differences in the two classes of receptors. Therefore, myasthenic sera may provide a valuable means to distinguish molecular species of Ach receptors.

The ability of Ach receptor proteins to undergo phosphorylation and dephosphorylation suggests they may contain protein kinase and phosphatase activity (Gordon and Diamond, 1980). Smilowitz et al. (1981) found that the Ach-receptor-rich preparation of electric organ showed three- to sixfold stimulation of endogenous phosphorylation in the presence of calmodulin and calcium. This stimulation of phosphorylation was inhibited by trifluoroperazine (TFP), an antagonist of calmodulin. Further analysis on SDS/polyacrylamide gels indicated that three of the receptor subunits, namely, those of MW 65,000, 58,000, and 50,000, were phosphorylated. These studies on phosphorylation require confirmation of their physiological function (Conti-Tronconi and Raftery, 1982).

Apart from the general agreement that the 40,000 MW protein (α-subunits) carries the Ach agonist or antagonist binding sites, no certainty exists regarding the functional significance of the other three subunits of the Ach receptor complex. Detergent-solubilized receptors from the electroplaques of *Torpedo* have been incorporated into lipid vesicles (Anholt et al., 1981), and such reconstituted receptors manifest agonist-stimulated cation exchange, which is blocked by α-toxins and antagonists. Such properties do not require either dimerization of the 9S form of the Ach receptor monomer or interaction with the 43,000 MW peripheral protein in a reconstituted system (Popot et al., 1981).

By using techniques for cloning DNA, Noda et al. (1982) have determined the primary structure of the α-subunit of the receptor protein in *Torpedo californica*. It is composed of 461 amino acids; 24 amino acids at the amino terminal resemble the signal peptide that is thought to be involved in the translocation of the secretory and membrane proteins into the endoplasmic reticulum (Section 15.4).

7.3. MORPHOLOGICAL CORRELATES OF Ach RECEPTORS

It was previously mentioned that the Ach receptors are highly concentrated in the postsynaptic membrane in the electroplaques in
electric fish and at the top of junctional folds in the neuromuscular
junction to the extent that there is no other integral membrane protein in these locations. Ach receptors are also concentrated in the
region of hot spots in cultured muscle cells and to a lesser extent in
denervated adult skeletal muscle (Conti-Tronconi and Raftery, 1982;
reviewed by Tipnis and Malhotra, 1979, 1981). Through a variety of
techniques (e.g., electron microscopy of negatively stained receptor-rich membranes from electroplaques, autoradiography with [125]I-
α-BGT, horseradish peroxidase-labeled α-BGT, and freeze-fracture
and freeze-etching), it is now generally agreed that the intramembranous particles on the freeze-fractured postsynaptic membranes and
the extrajunctional sarcolemma are the components of the Ach receptors (Figs. 7.2 through 7.5). There are 10,000–15,000 particles/
μm^2 in electroplaques in *Torpedo;* in muscle the particles range from
1800–7500/μm^2. There are two α-BGT binding sites per receptor
monomer, and in the native membrane the receptors exist as monomers and dimers. There is a fairly good agreement between the
number of estimated Ach receptors and the intramembranous particles in the electroplaques on the assumption that most of the particles represent a dimeric form of the receptor. However, in the case

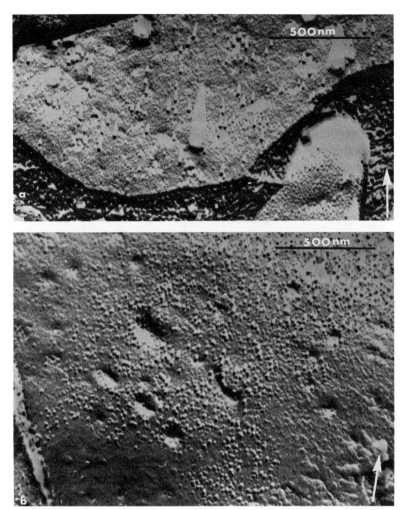

Figure 7.2 Freeze-fracture replicas of the rat EDL showing IMPs (~15 nm) on the sarcolemma in the region of neuromuscular junction (a) and comparable IMPs in the sarcolemma in the nonjunctional region after denervation of the muscle for 2 weeks (b). These 15 nm IMPs are most likely counterparts of the Ach receptors. In the normal adult muscle the nonjunctional sarcolemma does not show such large particles, which fits in with the lack of Ach receptors in the nonjunctional sarcolemma in the nondenervated muscle. Reproduced from Tipnis and Malhotra (1980).

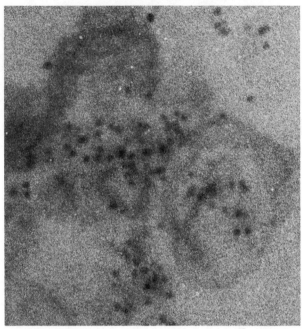

Figure 7.3 Electron micrograph of skeletal muscle membrane preparation in section aimed at localization of α-BGT binding sites (i.e., Ach receptors) by using ferritin-toxin conjugate. The dense granules are ferritin particles (~8 nm), indicating possible sites of toxin binding. The membranes are not well defined in the micrograph, presumably because of lack of adequate contrast. Sections were not stained with heavy metal salts. (Tipnis and Malhotra, 1979)

of muscle, there is a five- to tenfold discrepancy in the number of receptor molecules and intramembranous particles (see Fambrough, 1979). Either there are oligomers higher than dimers of receptor molecules in these membranes or some of the intramembranous particles are not revealed in freeze-fracture replicas because of high-angle shadowing during replication. Rotary shadowing might reveal better estimates of the number of intramembranous particles, and the enhancement of contrast by photographic rotation (Markham et al., 1963) might provide a more detailed image of the particles. In mammalian neuromuscular junction, a receptor complex may be

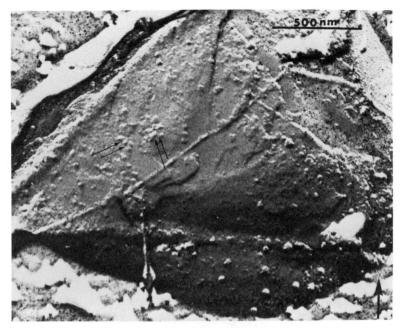

Figure 7.4 Electron micrograph of a freeze-etched membrane preparation from dener-vated extensor digitorum longus (EDL) muscle of the rat, incubated in ferritin-α-BGT conjugate (Tipnis and Malhotra, 1979). The membranes were frozen in nitrogen slush, fractured, and etched for 2 minutes. Double arrows indicate possible ferritin particles which are presumably sites of Ach receptors localized by binding with α-BGT. Running across the micrograph appears to be a collagen fiber on the surface of the membrane exposed by etching. Reproduced from Tipnis and Malhotra (1980).

composed of eight to nine receptor molecules organized around a central pit (Rash et al., 1978). In the electroplaques of *Torpedo*, the intramembranous particles occur in rosettes of three to six particles each (Cartaud et al., 1978).

7.4. STRUCTURAL MODELS

Based upon correlated electron microscopic and X-ray diffraction studies, Ross et al. (1977) proposed a model for the structure of

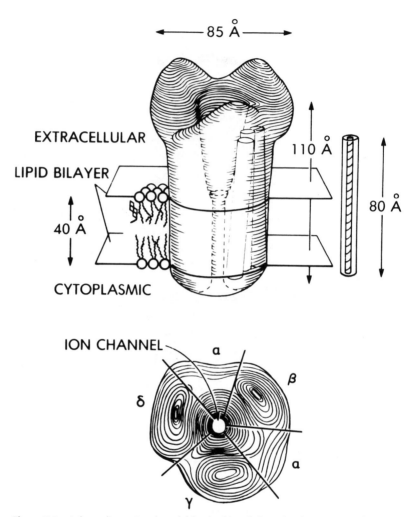

Figure 7.5 A three-dimensional model for the funnel-shaped Ach receptor molecule in the lipid bilayer. (Upper) On average 80Å long α-helices indicate an elongated shape of subunits that are arranged perpendicular to the membrane channel. The protein topography has been inferred from the densities in computer-filtered images of Ach receptor tubular lattices, from side views, and from X-ray diffraction. (Lower) Ach receptor molecule viewed from the synaptic side with borders arbitrarily drawn between elongated subunits and tentative assignment of subunit types consistent with our cross-linking data that show prominent α–γ and γ–δ linkages (disuccinimidyl tartarte, 6Å span). Karlin and colleagues (*Proc. Natl. Acad. Sci.*, U.S., in press) have found an average angle of 113° ±33° between the two α subunits per monomer, in agreement with our assignment. Reproduced with permission from Kistler et al. (1982).

membrane-bound Ach receptors. An electron density profile, derived from X-ray studies on Ach-receptor-rich membranes, at 6.5Å resolution across the membrane, displays the protein extending considerably more on one side (by ~5.5 nm) than on the other (~1.5 nm), total thickness of the asymmetric membrane being 11 nm. The lipid bilayer is ~4.0 ± 0.3 nm thick. High-angle diffraction patterns extend up to 17Å. These authors assigned a plane lattice of oblique symmetry (p1) for the arrangement of the Ach receptor. In this lattice there is only one molecule per unit cell. (A unit cell is based upon the unit cell vectors **a**, **b**, and **c**, and their magnitudes are designated as **a**, **b**, and **c**. These vectors are directed along the cartesian axes x, y, and z. Interaxial angles are described as α, β, and γ. For an oblique plane lattice, **a** ≠ **b** and $\gamma > 90°$.) Unit cell parameters were found to be **a** = 9.08 ± 0.41 nm, **c** = 9.10 ± 0.43 nm and $\beta = 118° ± 4$. Although the unit cell angles are close to 120°, both hexagonal and trigonal lattices have been ruled out because the amplitudes and the phases in the transform are not characteristically symmetric. (In a trigonal lattice, **a** = **b** = **c** and $\alpha = \beta = \gamma < 120°$, if the rhombohedral axes are chosen and **a** = **b** ≠ **c**, $\alpha = \beta = 90°$, $\gamma = 120°$ when referred to a hexagonal lattice [Henry and Lonsdale, 1965].) The Ach receptor oligomer has a MW ~370,000. The Ach-receptor-rich membranes give rise to a sharp reflection at 0.51 nm, oriented perpendicular to the membrane plane, which is indicative of α-helices in coiled conformation (Henderson, 1975). These α-helices appear to be ≈8 nm long and oriented perpendicular to the membrane surface. Another reflection noted at 0.63 nm poses some interesting possibilities. It can be accounted for by a number of possible structures (Ross et al., 1977):

(a) A coiled polypeptide of rather large diameter cross-linked as a parallel β-sheet structure.

(b) A twisted antiparallel β-sheet forming a barrel-like structure.

(c) A twisted antiparallel pleated sheet.

(d) Since such a 0.63 nm reflection has also been noted in the X-ray diffraction patterns of other membranes prepared from *Torpedo* and *Electrophorus*, including those enriched in acetylcholinesterase or ATPase, the likely possibility is raised that a 0.63

nm reflection arises because of some regular lipid structure in the membranes.

Of these possibilities, (a) and (b) can easily define an energetically stable channel of dimensions $\approx$0.5–1 nm (Ross et al., 1977). Since uranyl acetate, in negative staining, easily penetrates the center of each molecule of Ach receptor, it appears to be a plausible location for the presence of an ionophoretic channel. Klymkowsky and Stroud (1979) have recently labeled the Ach receptor molecules in the Ach-receptor-rich membranes by employing the antibodies raised against the Ach receptors and coupled to colloidal gold particles. Electron micrographs of such preparations have confirmed some of the conclusions of their previous work. Ach receptor molecules indeed extend above the extracellular surface by ~5.5 nm and little on the cytoplasmic side (see above).

Ach receptors visualized by electron microscopy of negatively stained folded-over vesicles reveal a funnel-shaped structure. The center of the receptor complex is stained more intensely with uranyl acetate relative to the membranous region between the receptor molecules. The stain seems to penetrate throughout the entire transmembrane length of the receptor (11.4 ± 1.9 nm) which is presumably the central ionic channel. A low-resolution three-dimensional model is seen in Figure 7.5, which displays the individual five subunits arranged around a central channel and provides an overall funnel-shaped form to the receptor (Kistler et al., 1982).

Brisson (1978) has reported tubular structures produced from Ach-receptor-rich membranes that exhibit a hexagonal type of packing. So far, published images of the Ach-receptor-enriched membranes do not have sufficient resolution to elucidate the symmetry of the Ach receptor molecule itself. However, optical filtering experiments suggest that the stain-excluding regions are arranged in a hexagonal pattern, though the constituting subunits in these regions were not resolved. Klymkowsky and Stroud (1979) suggested that perhaps Ach receptor particles have no internal symmetry, on the assumption that they are made up of four different subunits (see Hider, 1979).

Recently, Wise et al. (1979) reported a radius of gyration of 4.6 nm for the Ach receptor alone, based upon their analysis of the neutron-scattering results for the Ach receptor–Triton X-100 complex. An analysis of the extended scattering curve was not possible because of the low-coherent scattering and high-incoherent background. Nothing much can be said about the shape and structure of the Ach receptor because only a single parameter is known. The value of the radius of gyration and the molecular volume ($\sim 3 \times 10^5$ Å), however, suggest two alternative simple shapes, namely, oblate and prolate ellipsoids and cylinders. Alternatives about the shapes and sizes of the Ach receptor can be further narrowed down by incorporating the existing information on the structure of Ach receptors obtained from a variety of techniques such as electron microscopy, X-ray diffraction, and single-channel ionic conductance measurements. The most consistent model that Wise et al. (1979) have proposed consists of three stacked concentric cylinders and a cylindrical pore running through them. The pore has a diameter of 3 nm within the top cylinder and then narrows down to 1 nm through the middle and bottom cylinders.

Zingsheim et al. (1980) have obtained a projection of Ach-receptor-rich membranes from *Torpedo* at a resolution of 1.8 nm. The negatively stained membranes were examined with a scanning transmission electron microscope by using a low electron dose (<10 e/Å, which is 1/10 of the critical dose for most negatively stained biological specimen). The data were computed by a digital imaging process in which an average over many Ach receptor particles was obtained. The projection indicates that the Ach receptor protein extends into the aqueous medium and is an asymmetric structure. At the resolution of this projection (i.e., 1.8 nm) individual subunits could not be resolved, though a threefold symmetry is suggested. Zingsheim and coworkers therefore suggest that the Ach receptor protein molecule is made up of at least three (probably different) structural regions. The amino acid sequence of α-subunits derived from cloned DNA and analyzed by computer program for a possible secondary structure indicates that there is a long (64 amino acid residues) α-helical segment and several segments of possible α-helix and/or β-sheet.

The long α-helical segment contains a large hydrophobic, possibly a transmembrane, segment (Noda et al., 1982).

7.5. MOBILITY OF Ach RECEPTORS

In the adult neuromuscular junction and the electroplaques, the Ach receptors are packed at a very high density and do not manifest lateral mobility. Such an immobility could be due to receptor-receptor interaction as in the formation of dimers and/or interaction with a peripheral protein(s) such as the 43,000 MW protein associated with the receptor complex (reviewed by Conti-Tronconi and Raftery, 1982; see below). The formation of clusters of Ach receptors (hot spots) in myotubes and in denervated mammalian muscle could also be explained on similar grounds. Diffusely scattered Ach receptors, besides the hot spots, also exist in the myotubes and in the extra-junctional sarcolemma of denervated muscle. It was thought that the diffusely dispersed receptors are mobile, whereas the clustered receptors are immobile (Axelrod et al., 1976). However, more recent studies on the cultured primary myotubes indicate that the diffusely distributed receptors can become clustered and that the cluster formation takes place by trapping receptors as they diffuse in from the surrounding membrane. The formation of receptor clusters is inhibited by colchicine which disrupts microtubules. Stya and Axelrod (1983) made these observations on cultured myotubes by photobleaching with laser beam rhodamine-labeled α-BGT binding sites in a circumscribed region that contained clusters of receptors. After several hours (6.5–7.5 h), the clusters reappeared in the bleached region, which could have taken place by the aggregation of unbleached diffusely distributed receptors. While the role of microtubules in cluster formation is not understood, it appears that microtubules do not contribute to the stability of the clusters since existing clusters are not dispersed when cultured myotubes are treated with colchicine (Bloch, 1979). There does, however, appear to be differences in the "anchoring" of the Ach receptors in the clusters and the diffusely distributed receptors. As the clustered α-BGT-labeled Ach receptors in the myotubes were retained, presumably on the skeletal framework when the myotubes were extracted with a mild detergent Triton X-100, the diffusely distributed receptors were extracted, though partially, under the same conditions. It would therefore appear that the clustered Ach receptors are more tightly bound to the skeletal framework than the unclustered ones in myotubes (Prives et al., 1982).

In view of the high-density clustering of the Ach receptors at the neuromuscular junction (and electroplaques), it is most likely that in the adult, clusters are in a metabolic steady state. The turnover of the receptors then was found to take place by the addition of new receptors at the periphery of the clusters so that the older receptors were at the center of the cluster. Weinberg et al. (1981) determined such a mode of receptor turnover by prelabeling with α-BGT to block the existing Ach receptors in adult rat soleus muscle, and then one or two days later treating with ^{125}I-α-BGT to label newly inserted receptors. In autoradiography, receptor clusters appeared as annuli or doughnuts, such a distribution was confirmed by analyzing the radial silver grain density. The receptor removal takes place presumably by endocytosis, leading to the degradation of the receptors in lysosomes (Devreotes and Fambrough, 1976). The Ach receptors do not appear to be recycled, unlike the LDL receptors, which are recycled (Chapter 14).

7.6. FACTORS IN Ach RECEPTOR AGGREGATION

Factors that regulate the appearance of extrajunctional Ach receptors are not well understood and several suggestions have been made in the existing literature. Ach (nonquantal) has been proposed as one of the "neurotropic factors" regulating the appearance of extrajunctional Ach receptor. Blockage of cholinergic nerve transmission by infusion of α-BGT results in an increase in the extrajunctional receptors in the rat skeletal muscle comparable to that produced by surgical denervation. A similar, but less pronounced, effect is produced by blockage of the nerve transmission conduction by TTX (Drachman et al., 1982). A number of extracts from the nervous tissue have been reported to result in an increase in the Ach receptors when applied to muscle culture (Kalcheim et al., 1982). One such factor is a protein or a mixture of proteins of MW 100,000 (Podleski et al., 1978) and another is a small molecule of MW 2000 (Jessell et al., 1979). Yet another factor produced by cholinergic neuroblastomaglioma hybrid has been found to increase clusters of Ach receptors on myotube cultures from mouse, rat, and chick (Christian et al., 1978). (This last-mentioned factor may be an "aggregation" factor that acts by aggregating dispersed mobile receptors; the phenomenon is thus comparable to the lectin-immunoglobulin-induced cap formation or patch formation.) The neurotrophic factor involved in the formation of Ach receptor clusters could be simulated by polylysine-coated latex beads (Peng et al., 1981). The formation of Ach receptor clusters could be induced in pure muscle cell cultures by incubation with positively charged latex beads for 1–2 days. The cluster formation was associated with the site of contact with the bead, there being no evidence from scanning electron microscopy that the beads were phagocytosed. These observations suggest that the Ach receptor cluster formation in the muscle is induced by a surface interaction between the muscle and an exogenous agent that in vivo may be released from the nerve. The incubation with the beads suppressed preexisting Ach receptor clusters in muscle in the area where no beads were detected. Such a phenomenon may be operational during the initial stages of synaptogenesis.

Besides neurotrophic factors, intracellular calcium has also been implicated in the regulation of synthesis of Ach receptors (see McManaman et al., 1981). In the cultured myotubes from rat, the Ach receptors were decreased by 25–30% when the myotubes were incubated in a calcium-deficient medium for 24 hours. It is known that the contractile activity of the muscle involves fluctuation in the myoplasmic calcium level, and muscle activity influences the level of Ach receptors. McManaman et al. (1981) proposed that the regulation of Ach receptor is influenced by the calcium level in a manner not yet understood. It does not appear to be related in a simple reciprocal fashion, that is, increased Ach receptor levels associated with a decreased calcium level and decreased receptor level with an increased calcium level in the myoplasm. Cyclic nucleotides have also been implicated in the control of biosynthesis of Ach receptors, and this control could operate indirectly through calcium (Betz and Changeux, 1979).

It is apparent that muscle activity and electrical activity of the nerve exercise regulatory control of the biosynthesis and expression of Ach receptor activity in the sarcolemma. Calcium, Ach, and other neurotrophic factors may regulate the formation of the neuromuscular junction and the clustering of Ach receptors. Such factors could serve multiple roles, as Miledi and Uchitel (1981) found in the frog. Slow muscle fibers of the frog that are normally incapable of generating action potentials do so after an intramuscular injection of α-BGT. α-BGT blocks Ach receptors and thus interferes with the action of nonquantal acetylcholine leaking from the presynaptic ending. (It is also conceivable that α-BGT has other as yet unknown effects on nerve and muscle.) However, in the slow muscle, the ability to generate action potentials is normally repressed, and this is thought to result from the failure of incorporation of sodium channels in the membrane of the slow fiber: The prime candidate involved in the control of the synthesis and incorporation of these channels could be the Ach (nonquantal) that leaks from the presynaptic ending.

7.7. PERIPHERAL PROTEIN ASSOCIATED WITH Ach RECEPTORS (43,000 MW PROTEIN)

Associated with the Ach receptor complex is a protein of MW 43,000 which falls into the category of peripheral proteins (see Barrantes, 1982; Changeux and Dennis, 1982) and appears to participate in protein–protein interactions with the receptor subunits. It is possibly involved in the stabilization of dimeric species (13S) of the Ach receptor by cross-linking through disulphide linkages. On the basis of immunofluorescence studies on the *Torpedo*'s Ach-receptor-rich membranes which had been made permeable by saponin treatment, Barrantes (1982) observed that a nonreceptor ν-peptide (MW 43,000) is predominantly exposed on the cytoplasmic surface. Similar results have been reported from several laboratories. Antiserum raised by immunization with the alkaline-extracted proteins from the highly purified Ach receptor membranes from *Torpedo* react with the 43,000 MW protein. (No reaction was detected with any of the four subunits of the receptor protein.) Immunofluorescence staining showed intense staining at the innervated surface of the electroplaques but not at the uninnervated surface (Froehner et al., 1981).

Sealock (1982) demonstrated by using tannic acid as a negative stain for electron microscopy and biochemical approaches that the 43,000 MW protein is a peripheral protein in the Ach-receptor-rich membranes of the electric organ of the electric fish, which could be extracted by incubation at pH 11—a procedure that removes peripheral protein. This peripheral protein extends ~4.5 nm beyond the plasma membrane into the cytoplasm, whereas the Ach receptor protein extended 4–5 nm into the extracellular surface.

An interaction between the Ach receptor and the 43,000 MW protein is inferred from observing that after alkaline extraction, the sensitivity of the receptor to trypsin attack and to heat denaturation increases and the receptor becomes mobile.

St. John et al. (1982) came to the same conclusion as Sealock (1982). They used isolated *Torpedo* membranes and studied iodination of proteins in intact vesicles (mostly extracellular side-out vesicles) which were impermeable to ferritin or lactoperoxidase, but became permeable after saponin treatment. In the latter case, the 43,000 MW protein became available for iodination, thereby indicating its accessibility to the cytoplasmic surface of the plasma membrane. Cartaud et al. (1981) reported that alkaline treatment of the Ach-receptor-rich membrane of *Torpedo* removed this peripheral protein of 43,000 MW. Coincident with the removal, the densely staining coat on the cytoplasmic face of the postsynaptic membrane also disappeared. It has been suggested that this protein interacts with the membrane and is responsible for immobilization of the Ach receptors in the postsynaptic membrane. Alkaline extraction of this protein enables the Ach receptors to become mobile since the freeze-fractured and freeze-etched faces showed redistribution of the macromolecular assemblies thought to represent the Ach receptors.

The 43,000 MW protein is most likely also present in the mammalian skeletal muscle Ach receptor complex. Antiprotein 43,000 MW serum raised against *Torpedo* stained equally both *Torpedo* and rat diaphragm muscle in cyrostat sections, and the staining was restricted to the synaptic regions. The result is consistent with the suggestion that this antigenic protein is conserved through evolution (Froehner et al., 1981).

7.8. Ach RECEPTORS IN CNS AND SYMPATHETIC NEURONS

The structure of the nicotinic Ach receptor of the central nervous system (CNS) is of current interest because α-BGT has been reported to fail to block cholinergic transmission in the CNS, while capable of such a blockage in muscle. Norman et al. (1982) purified α-BGT binding protein from the chick optic lobe and from the rest of the brain. This receptor from the CNS has a value of 9.1S, as does the skeletal muscle receptor, but it differs from the latter in that there is a single subunit of MW 54,000. The research revealed only a weak cross-reactivity between the optic lobe receptor and the antisera against chick skeletal muscle Ach receptors. Apparently there are weak immunogenic determinants that are common to the Ach receptors in the CNS and peripheral nervous system in the chick. There may be other differences between the Ach receptors in the two locations. For example, apart from the inabilty to demonstrate blockage of cholinergic neurotransmission by α-BGT in CNS, there is a marked decrease in α-BGT binding sites after surgical denervation. The skeletal muscle is well known to undergo Ach supersensitivity resulting from incorporation from newly synthesized receptors into the extrajunctional sarcolemma (Section 7.1).

In contrast to the above findings on the chick brain, the goldfish brain α-BGT binding sites are distinctly comparable to those of the muscle Ach receptors (Oswald and Freeman, 1979): an isoelectric focusing point of ~5.0, S value 11.45, MW 340,000 for the toxin-receptor complex, with an estimate of ~8 nm for the size of the complex. α-BGT also blocked synaptic transmission to the goldfish optic tectum.

It appears that α-BGT may not be a valuable probe for characterization of Ach receptors on different cell types. α-BGT binding sites also exist in sympathetic neurons and a clonal sympathetic nerve cell line (PC 12), but α-BGT does not block agonist-induced activation of Ach receptors.

Another α-neurotoxin other than α-BGT that failed to block transmission in the chicken ciliary ganglion was effective in the sympathetic ganglion of the bullfrog. α-BGT binds specifically to rat and chicken autonomic neurons but does not block Ach receptor activation even at saturating concentrations ($\sim$10 nM). (The possibility therefore arises that the ability of α-BGT to block synaptic transmission reflects phylogenetic differences in neuronal Ach receptors between the sympathetic ganglion of bullfrog and of higher vertebrates [Marshall, 1981].) Another snake venom toxin, α-mambatoxin, which binds to muscle and PC 12 cells and blocks α-BGT binding, may be more useful for comparison of Ach receptors. α-mambatoxin binds to the Ach receptor in muscle, which can be recognized by immunoprecipitation, but in the PC 12 cells the α-mambatoxin binding component is not recognized by anti-Ach receptor antibodies, which do not recognize Ach receptor in these cells (Patrick et al., 1980). For yet unknown reasons, the number of α-mambatoxin binding sites is twice that of α-BGT binding sites in both nerve and muscle in situ, but in detergent solubilized muscle cells the number of α-mambatoxin is reduced in half to equal the α-BGT binding sites.

Cryptic α-binding sites may exist in some cases—Rehm and Betz (1981) have reported that phospholipase c treatment of retinal membranes from newly hatched chicken resulted in a twofold increase in ^{125}I-α-BGT binding to the membranes. These unmasked α-BGT binding sites were indistinguishable in their pharmacological characteristics from the control membranes. However, phospholipase-c-treated membranes showed that one-half of the toxin binding sites were inhibited by CO^{2+} (greater than 1 mM), whereas the controls were insensitive up to 8 nM CO^{2+}. The origin of the cryptic α-BGT binding sites is not understood. However, it is conceivable that the cryptic and the accessible α-BGT binding sites are located on the same Ach receptor molecule since there are two known agonist binding sites in muscle and fish electric organs. Embryonic chick pectoralis muscle showed only a moderate increase ($\sim$30%) in α-

BGT binding sites after treatment with phospholipase c. These differences in muscle and retina may simply reflect differences in the fractional occupancy of their α-BGT binding sites by the toxin (Section 7.1).

It is apparent from the above that the structure of the Ach receptors in the CNS and sympathetic neurons is much less clear than that of their counterparts in muscle and electroplaques. However, the techniques of cloning DNA are now being applied to the Ach receptor protein (Barnard and Dolly, 1982; see Stevens, 1982) which should facilitate advances in understanding the relationship between the receptor proteins in the CNS and the peripheral nervous system. Genetical manipulations of the receptor protein should provide information on the functioning of the receptor complex.

8

Acetylcholinesterase (AchE)

AchE occurs in various molecular forms, but the high-molecular-weight form referred to as the 16S or 18S form is of particular interest to neurobiologists. It is highly concentrated in the region of the neuromuscular junction and the electric organ of the fish, where it functions in the termination of impulse transmission by hydrolysis of the neurotransmitter Ach (Marnay and Nachmansohn, 1938). It can easily be demonstrated by histochemical techniques, both at light microscopical as well as at electron microscopical levels (Figs. 8.1 and 8.2; Malhotra and Tipnis, 1978; Tewari et al., 1982). As for the biosynthesis of AchE, which is especially interesting, it appears that both nerves and muscle may contribute to the expression of this enzymatic activity in the synaptic cleft. This tentative conclusion is based upon recent findings. Silberstein et al. (1982) have reported that the 16S form of AchE is expressed on the surface of cells of a mouse muscle line cultured in the absence of nerves. However, this particular cell line (C_2) is derived from adult muscle and may represent an advanced stage of myoblast differentiation when in vivo neuronal influences have already been expressed. Similarly, that

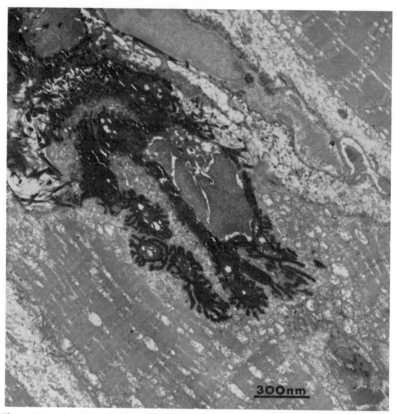

Figure 8.1 Histochemical localization of acetylcholinesterase (AchE) in extensor digitorum longus (EDL) of the rat. The substrate was acetylthiocholine iodide and the procedure was that of Karnovsky and Roots (see Malhotra and Tipnis, 1978). The reaction product is highly concentrated in the region of the neuromuscular junction and contains copper, iron, iodine, and sulphur (see Fig. 8.2).

muscle cells from older embryos are able to synthesize the 16S form in the absence of nerves has been demonstrated clearly by Koenig and Vigny (1978), whereas in the muscle cells from younger embryos, neural induction was found to be necessary for the biosynthesis of the 16S form of AchE. Inestrosa et al. (1981) reported the presence of the 16S form of AchE on the surface of a nerve-like cell line pheochromocytoma PC12, which raises the possibility that this

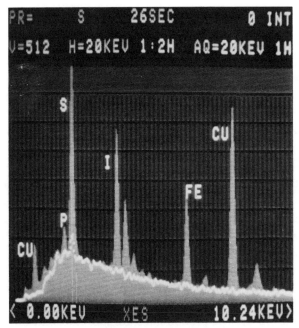

Figure 8.2 X-ray spectrum of the AchE reaction product in the EDL muscle (shown in Fig. 8.1) superimposed on the background (white line). Copper, iron, iodine, and sulphur are well above the background. The details of this investigation appear in Tewari et al. (1982).

form of AchE could be at least in part contributed by the neuron at the neuromuscular junction. Such an interpretation is consistent with the findings that the AchE decreases at the neuromuscular junctional site after denervation of the rat muscle (Hall, 1973; Malhotra and Tipnis, 1978). Blockage of axonal flow also decreases the AchE at the neuromuscular junction (Fernandez and Inestrosa, 1976). Whereas the 16S form of AchE in the rat undergoes a marked reduction following denervation, the rabbit slow-twitch muscle manifests an increase when denervated, thereby indicating that the regulation of varying levels of AchE is dependent upon the species and the type of muscle (Bacou et al., 1982).

Notwithstanding the role of nerve and muscle cells in the biosynthesis of AchE, McMahan et al. (1978) determined that at least some of the enzymatic activity is associated with the basement membrane (basal lamina), a finding of considerable value since the basal lamina is commonly considered to be a structural component of tissues. McMahan (1978) and coworkers demonstrated such an association in the neuromuscular junction of the frog. They denervated frog by removing a long segment of the nerve to the cutaneous pectoris muscle, which produced degeneration of the nerve terminals. They also disrupted muscle by mechanical injury and irradiated frogs with X-rays to prevent regeneration of new muscles into the basal lamina. Histochemical studies showed that a specific AchE activity was present in the basal lamina in the region identifiable by the junctional folds. These findings clearly indicate that most, if not all, of the AchE is tightly associated with the basal lamina in the neuromuscular junction. It is of interest that an enzyme was reported for the first time to be bound to the basal lamina of a cell.

AchE can be readily dissociated from the membranes of the electric organ and neuromuscular junction by mild protease treatment. Massoulié (1980) has reviewed the structure and polymorphism of the AchE. The predominant form of the AchE in the electric organ of the fish is the tailed-form of the enzyme, which has been reported to exist as one (9S), two (14S), or three (18S) tetrameric assemblies of catalytic subunits and a rod-like tail ~50 nm long (Dudai et al., 1973). The tail is a collagen-like triple helix in which the peptides are covalently linked to each other and through disulfide bonds to the catalytic subunits (Anglister and Silman, 1978). This tailed form of AchE has an essentially similar structure in the electric organs of *Torpedo* and *Electrophorus*. A similar form of AchE (16S) also exists at the neuromuscular junction of the rat and other higher vertebrates. However, in higher vertebrates, the tailed form of the AchE constitutes only a small fraction of the total AchE, and in these vertebrates the major fraction is made up of globular forms of AchE which occurs again in monomeric, dimeric, or tetrameric form (Fig. 8.3).

AchE has been isolated and investigated extensively from the electric organ of the electric fish and mammalian neuromuscular junction. In addition, several molecular forms of AchE with varying S values have been reported from different invertebrate and vertebrate tissues. For example, four molecular forms (5.4S, 7.1S, 11.5S, and 19.5S) have been reported from chick embryo leg muscle but these are assemblies of a common monomer. The relative abundance of these forms may change during development (Rotundo and Fambrough, 1979).

AchE presence in the perikarya and axons of neurons may be related to the release of the amount of the transmitter, so that motor neurons of high AchE activity may innervate fast muscles, whereas nerves with low AchE activity innervate slow muscle nerve fibers (Gruber and Zenker, 1978). The glial cells of the superior cervical ganglion in the rat begin to synthesize AchE (4S and 10S form) after denervation, whereas no such activity was detectable in the normal glial cells (Gisiger et al., 1978).

Besides its well-known above-mentioned role in the hydrolysis of the neurotransmitter Ach, AchE serves in uncleaved eggs and during the early phase of development as a biochemical marker for cell differentiation. For example, Fitzpatrick-McElligott and Stent (1981) followed the distribution of AchE activity (by using a histochemical procedure) from the uncleaved eggs of leach to the early stages of embryogenesis, and found that the presence of AchE activ-

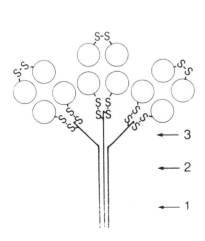

Figure 8.3 Structure of the collagen-tailed (A_{12}) molecule, from the data of Rosenberry and Richardson, Anglister, and Silman and Bon and Massoulié obtained from *Electrophorus* enzyme. The disulfide bonds that link the catalytic subunits as dimers to the tail strands are shown schematically. Other disulfide bonds exist within each catalytic subunit and between the tail peptides. The points of cleavage by collagenase are indicated by arrows: position 1 is sensitive to collagenase at 20°C, and positions 2 and 3 at 37°C. Trypsin probably digests the entire tail, dissociating active tetramers. Trypsin and collagenase (at 37°C) solubilize the enzyme as the tetrameric lytic form (G_4). Reproduced with permission from Massoulié (1980).

ity precedes the formation of any nervous tissue. The role of AchE in the development prior to the formation of the nervous system remains to be investigated.

There is a soluble form of AchE whose concentration in the amniotic fluid has been found to be higher in cases of neural abnormality than in normal amniotic fluid. The possibility has been suggested that an analysis of the AchE in amniotic fluid may turn out to be of value in antenatal diagnosis of neural tube defects, if further studies confirm a general correlation between high levels of AchE and neural defects in embryos (Chubb, 1980).

A great deal of homology may exist between AchE in different cellular locations, as evidenced by the fact that five monoclonal antibodies raised against human erythrocyte AchE cross-reacted with AchE in the neuromuscular junction of man and monkey (Fambrough et al., 1982).

9

Purple Membrane of Halobacteria: Bacteriorhodopsin

Purple membrane of halobacteria (*Halobacterium halobium*; Stoeckenius, 1976) constitutes one of the best-understood membrane systems in the respect of correlation of structure and function. Purple membranes are patches in the cell membrane in halobacteria that live in an environment containing very high NaCl concentrations, near saturation. The remaining part of the cell membrane is referred to as the red membranes, which contain the components of the respiratory chain and serve in oxidative phosphorylation (Stoeckenius, 1976). The purple membranes (thickness 4.5–5.0 nm; Blaurock and Stoeckenius, 1971) contain a single protein, bacteriorhodopsin (MW 26,000) that makes up 75% of the membrane mass, the remaining 25% being lipid. The lipid composition is almost identical to that in the rest of the cell membrane except for the main carotenoids, which are absent. Phosphatidylglycerosulfate and glycolipid sulfate are exclusively localized in the purple membrane. All lipids are derived from isoprene. Phospholipids are diether analogues of phosphatidylglycerophosphate; the rest of the polar lipids are mainly glycolipids. The hydrocarbon moiety in all polar lipids is comprised of dihydrophytol chains ether-linked to glycerol (Stoeckenius, 1980) which have saturated branched fatty acid chains, as compared to the lipids from membranes of other organisms, which have unbranched fatty acid chains.

The purple color is caused by the chromophore retinal (vitamin A aldehyde) that is bound as a protonated Schiff base to the ε-amino group of lysine residue. This protein is strikingly similar to the rhodopsin in the retina of higher animals and functions in photoreception (Oesterhelt and Stoeckenius, 1971; Stoeckenius, 1980). The purple membranes can easily be isolated since they remain intact in low ionic strength solutions, whereas the rest of the membrane dissociates. They may make up to 50% of the total cell membrane. The protein is organized in a two-dimensional hexagonal lattice, which makes the purple membranes well-suited for high-resolution structural studies.

By using an ingenious procedure for the analysis of periodic arrays of biomacromolecules in which data from low-dose electron images and diffraction patterns were combined, Henderson and Unwin (1975) reconstructed a three-dimensional image of purple membranes at 0.7 nm resolution. Glucose embedding was employed to alleviate the dehydration damage and low doses ($<0.5e/Å^2$) to reduce the irradiation damage. The electron micrographs of unstained membranes were recorded such that the only source of contrast was a weak phase contrast induced by defocussing. In these extremely noisy featureless micrographs, the weak signal was retrieved from noise by averaging over thousands of unit cells by means of Fourier transforms. Phases were obtained from the computed transforms after suitable correction for contrast transfer function, while the electron diffraction patterns provided the amplitudes of the structure factors. By employing data obtained for the various tilts of the membrane, a three-dimensional Fourier synthesis was computed which, strictly speaking, gave a map of the electrostatic potential in the unit cell (Henderson and Unwin, 1975). In the map, the globular

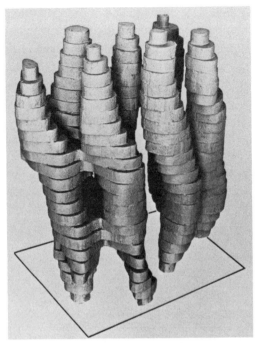

Figure 9.1 A model of a single protein molecule in the purple membrane, viewed roughly parallel to the plane of the membrane. The top and bottom of the model correspond to the parts of the protein in contact with the solvent, the rest being in contact with lipid. The most strongly tilted α-helices are in the foreground. Reproduced with permission from Henderson and Unwin (1975).

protein extends to both sides of the lipid bilayer and is comprised of seven α-helices packed about 1–1.2 nm apart and 3.5–4.0 nm in length, running perpendicular to the plane of the membrane (Fig. 9.1). The molecules are organized around a threefold axis with a 2-nm-wide space at the center that is filled with lipids. This elegant work represents the most significant step forward thus far, as it has for the first time provided us with the structure of an integral membrane protein in situ. Further work on purple membrane by low-dose electron microscopy has involved the determination of the orientation of the resulting structure with respect to the cell (Hayward et al., 1978; Henderson et al., 1978). The confirmation of the monomer boundary was done as a consequence of the structure determination of the orthorhombic form of purple membrane (Michel et al., 1980). By exploiting the reduction of radiation damage at liquid nitrogen temperature, Hayward and Stroud (1981) recently obtained the projected structure of purple membrane at 3.7 nm resolution. However, the figure of merit which reflects the precision in phase determination dropped below 0.5 at the outer resolution limit. (The "m" or the figure of merit is the mean value of the cosine of the error in phase angle for the reflection. A value of $m = 1$ reflects zero error in phase angle, a value of $m = 0.74$ indicates approximately $42°$ error, and a value of $m = 0.24$ corresponds to an error of $76°$. For a discussion see Blundell and Johnson, 1976.) Extension of high resolution three-dimensional phasing from the tilted electron images poses a severe limitation (Hayward and Stroud, 1981).

Henderson et al. (1982) have recently reported on the projected structure of a modified trigonal form of purple membrane with a unit cell dimension of about 5.9 nm instead of the normal 6.27 nm and having entirely different diffraction intensities. Analysis of different packing arrangements would reveal not only how the molecules interact in different environments, which is of relevance to model building (Engleman et al., 1980), but could also be employed in the determination of high-resolution three-dimensional phases by the method of molecular replacement (Rossmann, 1972). Very recently, Rossmann and Henderson (1982) used the molecular replacement method for the phasing of electron diffraction amplitudes and extended the phases beyond the current 6Å resolution limit (posed by the irradiation damage and inherent lack of long-range order within the specimen, in addition to instrumental perturbations) to obtain the projected structure of purple membrane at 3.3Å resolution. Even in this projection, since the 45Å thick helices are being viewed end-on, there is no significant gain in the structural information. A certain degree of flattening is apparent in the center of three helices, 5, 6, and 7 (Engleman et al., 1980), as well as some extra detail in the unresolved helices 1, 2, 3, and 4. Certain features in the projection may perhaps correspond to the positioning of hydrocarbon chains or ordered lipid. Using constrained density map modification and refinement methods, Agard and Stroud (1982) have been able to minimize the distortions caused by the missing cone of electron microscopic data in the three-dimensional model of Henderson and Unwin (1975). The density in the refined maps displays at least five extrahelical connectives at both the cytoplasmic and extracellular surface of the membrane. The bacteriorhodopsin molecule appears to be 45Å thick, and the helices extend untapered to both membrane surfaces. These clues have led the authors to propose the five most probable models, two of which are similar to the ones earlier proposed by Engleman et al. (1980). Since the exact initiation and termination of helices is still not clear from the map, further discrimination among various models based solely on the scattering density map would be premature.

The amino acid sequence of bacteriorhodopsin has been finally unravelled after years of effort to determine it (Khorana et al., 1979; Ovchinnikov et al., 1979). The availability of the amino acid sequence, together with information about the electron scattering density from the work of Henderson and Unwin (1975), has stimulated the model-building efforts (Engleman et al., 1980) to fit the bacteriorhodopsin sequence information into a series of α-helical segments. Some of the features of the model proposed by Engleman et al. (1980, 1982; Fig. 9.2) are experimentally validated in Engleman and Zaccai (1980). Employing difference Fourier techniques in neutron scattering on native and modified (with all the valines or phenylalanines deuterated) purple membranes, they mapped out the distribution of valines and phenylalanines in the projections of purple membrane structure. Valine was shown to be distributed toward the periphery, whereas the phenylalanine was distributed toward the center of a single bacteriorhodopsin molecule. It appears as if the charged and polar groups of the bacteriorhodopsin molecule lie in the interior of the molecule away from the lipids, whereas the nonpolar groups are exposed to the exterior in contact with the lipid molecules. In comparison to the "oil-drop" organization of water soluble proteins (where the charges or polar groups are exposed to the aqueous solvent and the nonpolar groups are clustered in the protein interior), the bacteriorhodopsin appears to have an "inside-out" structure. Extensive neutron diffraction studies by Rogan and Zaccai (1981) and Zaccai and Gilmore (1979), employing H_2O-D_2O

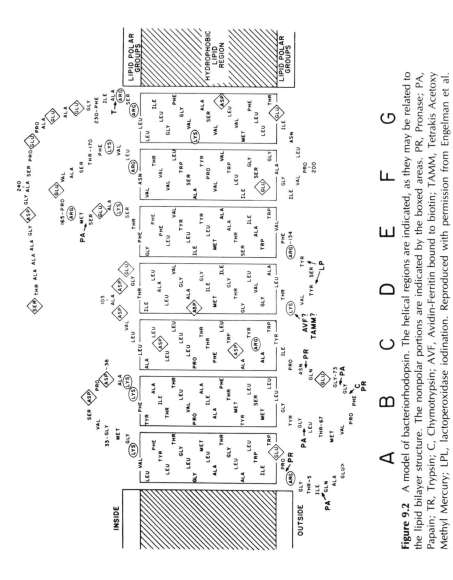

Figure 9.2 A model of bacteriorhodopsin. The helical regions are indicated, as they may be related to the lipid bilayer structure. The nonpolar portions are indicated by the boxed areas. PR, Pronase; PA, Papain; TR, Trypsin; C, Chymotrypsin; AVF, Avidin-Ferritin bound to biotin; TAMM, Tetrakis Acetoxy Methyl Mercury; LPL, lactoperoxidase iodination. Reproduced with permission from Engelman et al. (1982).

96

exchange, have eliminated the possibility of any substantial aqueous channel in the protein.

Spatial location and orientation of the retinal chromophore remain to be established. The chromophore appears to be very tightly bound to the protein, based upon transient absorption dichroism measurements, which revealed a complete absence of rotational relaxation in a time period of less than 10^{-2} seconds (Kouyama et al., 1981; Razi-Naqvi et al., 1973). The chromophore is tilted slightly out of the membrane plane. By employing various techniques the value of tilt angle appears to be between 66 and $\sim 71°$ with respect to the plane normal to the membrane (Kimura et al., 1981; see Stoeckenius, 1980).

In a detailed investigation based upon the quantitative analysis of fluorescence energy transfer, Kouyama et al. (1981) have determined the most probable location of the chromophore in the plane of the membrane. The main body of the retinal chromophore appears to be buried in a rigid pocket within a single protein surrounded by four to five α-helical regions in such an orientation that the consequent dipole–dipole interaction with the neighboring chromophores are reduced to a minimum. Neutron diffraction studies of purple membranes with ^{1}H-retinal and ^{2}H-retinal have revealed that the retinal is located 1.7 nm below the membrane surface (King et al., 1979). In a subsequent study, King et al. (1980) demonstrated that the β-ionone ring portion of the retinal is situated between the α-helical segments, 2.6 nm away from its nearest neighbor.

[The only other known naturally-occurring membrane systems which have so far lent themselves to three-dimensional image reconstructure are the gap junctions (Section 6.3) and the Ach-receptor-rich membrane of the electric ray *Torpedo californica* (Section 7.4). Both these membrane systems have been mapped to a resolution of 1.8 nm. The gap junction protein, connexin, is arranged in a cylinder composed of six subunits that surround a relatively large, ~ 2 nm in diameter, channel. Ions and molecules up to ~ 1000 MW can pass through this channel (Unwin and Zampighi, 1980). The Ach receptor complex, which consists of four polypeptides $(2:1:1:1)$ arranged around an ionophoretic transmembrane channel, has been shown to have an asymmetric structure (Zingsheim et al., 1980). The channel appears like a funnel, wider at the top (~3 nm) and narrowing to ~1 nm in the middle and on the cytoplasmic side (Wise et al., 1979). This channel is nonselective since both Na^+ and K^+ pass through it, in opposite directions, when it is opened by acetylcholine. It should be of interest to compare the structure of these transport proteins when further details of the structure of the connexon and Ach receptor protein emerge.]

In the presence of O_2, ATP is synthesized by oxidative phosphorylation; when O_2 is deficient halobacteria turn to a photosynthetic mode in which bacteriorhodopsin plays a role in producing ATP. Bacteriorhodopsin functions in the light-driven proton pump in which the protons are taken up on the cytoplasmic side of the purple membrane and released into the medium, thus producing a proton concentration gradient which generates an electrical potential across the cell membrane. This light-dependent electrochemical gradient can drive the synthesis of ATP from ADP and the protons flow back through the ATPase: the ATPase ejects protons when it hydrolyses ATP (Stoeckenius, 1976). The intermediates in this photoreaction are detailed by Stoeckenius (1980).

As mentioned above, the use of neutron diffraction on the purple membranes in which H_2O is exchanged with D_2O has ruled out the notion that a bulk water channel of 1 nm diameter or larger extends through the membrane (Zaccai and Gilmore, 1979). A transmembrane chain of hydrogen bonds provided by the hydrophilic groups of amino acid side-chains could conduct protons across the membrane (Ovchinnikov et al., 1979; Stoeckenius, 1980). Such a mechanism has been proposed for transport of protons across biological membranes (Nagle and Morowitz, 1978). This involves formation of a continuous chain of hydrogen bonds between the protein side-groups such as the hydroxyls of serine, threonine, and tyrosine and the carboxyls of aspartic and glutamic acids. A chain of 20–25 successive hydrogen bonds would suffice to traverse the biological membrane. Based upon studies on ice, the hydrogen bonded chain can rapidly transport protons from a region of high electrochemical potential across a membrane with little loss of energy. Before the high-energy protons are released into a solution with a low electrochemical potential, they could drive a chemical reaction, as in the synthesis of ATP by the generally accepted scheme of chemiosmotic theory (Mitchell, 1966). Nagle and Morowitz (1978) speculate that such a basic proton conduction mechanism may turn out to be operational in proton transport in bioenergetic processes.

Khorana et al. (1979) have reported the amino acid sequence of bacteriorhodopsin, containing 248 residues (which differs somewhat from the sequence reported by Ovchinnikov et al., 1979). What is apparent is the clustering of hydrophobic and hydrophilic amino acids in several tracts, there being 71.8% of hydrophobic and 28.2% hydrophilic amino acids. The COOH terminus is on the cytoplasmic side and NH_2 terminus on the exterior surface, and there are several residues that appear to extend out of the lipid bilayer. Earlier, Blaurock and King (1977) suggested on the basis of their X-ray diffraction data that the bacteriorhodopsin molecules may extend out of the lipid bilayer.

Estimates of the contents of IMPs in freeze-fracture replicas indicate that each IMP (10–12 nm) contains 9–12 molecules of bacteriorhodopsin, which adds up to 63–84 transmembrane α-helices (Fisher and Stoeckenius, 1977). In intact cells, the IMPs are arranged in a hexagonal pattern on the P face in the freeze-fractured replicas, whereas the complementary E face appears relatively smooth. However, in lipid vesicles in which bacteriorhodopsin was incorporated, the two fractured faces showed an opposite pattern of distribution of IMPs, as did the purple membrane vesicles made from the cells by lowering the salt concentration of the medium. These observations show that the membrane vesicles made by breaking open the cells are inside-out vesicles (Stoeckenius, 1976).

The presence of seven α-helices in bacteriorhodopsin, discussed above, has been questioned by recent data derived from UV-CD and IR spectroscopy of the purple membrane. The α-helical content appears to be sufficient for only five helices and, thus, leaving a substantial amount for as many as four strands of β-sheet (Jap et al., 1982).

Apart from the bacteriorhodopsin, the presence of another retinal pigment halorhodopsin, has been identified in the *Halobacterium halobium*. This retinal pigment has been reported to mediate a light-driven sodium pump in this microorganism (Lanyi, 1981).

10

Ion Transport Channels

10.1 INTRODUCTION

Lipid bilayer poses a strong electrostatic barrier to ion transport, and therefore special mechanisms have evolved to facilitate the permeation of ions and other hydrophilic substances across the lipid bilayer. Two mechanisms for such transport are known: carrier mechanism and the channel pathway. A carrier is usually defined as a transport device that is alternately exposed to one surface or another of the bilayer. Though ion carriers exist, they cannot transport Na and K ions at the known rates of $>10^6$ ions/s per channel (Edmonds, 1981). A channel must provide such an environment for the ion that its electrostatic self-energy during its passage is at least equal to its value for the fully hydrated ion (Edmonds, 1981). Based on theoretical considerations of the nature of electrostatic barriers posed by the lipid bilayer to ion transport, Edmonds (1981) proposed a general model for transport of ions through a channel (Fig. 10.1). In this model, the channel is conceived of as "a cage or net structure of water molecules composed largely of planar pentagon and puck-

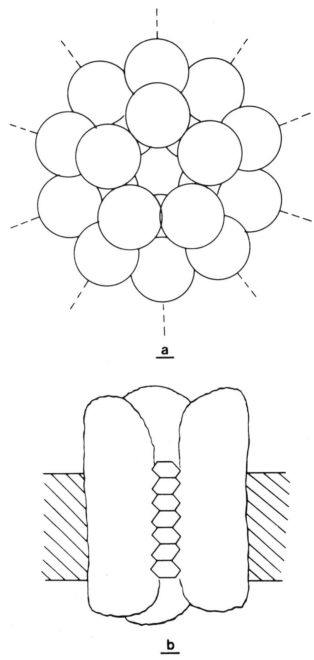

Figure 10.1 (a) A view down the z-axis of a dodecahedron of water molecules. The water molecules are represented by spheres 0.28 nm in diameter and the unsaturated

102

ered hexagon rings." A dodecahedron composed of pentagon rings is a particularly stable water structure, and a channel can be formed by stacking dodecahedra along a common z axis perpendicular to the membrane. The channel is supported by five helical rods that represent the channel protein(s). In this model, ion transfer is controlled by electric fields that exist in the channel because of electrical order. Selectivity is determined by the ionic binding site at the center of each pentagon ring. For example, a planar pentagon ring favors Na^+ over K^+, whereas a puckered hexagon favors K^+ over Na^+. The ions would move over the water channel structure (rather than through it), passing between adjacent pentagons having a shared edge.

Interestingly enough, the modeling of Na^+ and K^+ ion transfer rates as a function of membrane voltage differences based upon the water channel model displays rather good agreement with the experimentally determined rates. A feature unique to the electrically controlled water channel model is that it may accelerate or greatly retard the ion transfer but is never fully closed (whereas the mechanically opened and closed models can fully close). This feature manifests itself at the extremely negative membrane voltages (when the mechanical models are fully blocked). However, there is no clear or direct structural evidence for the existence of ordered water channels in membranes. Yet, some of the structural features of Ach-activated postsynaptic channels appear to be consonant with the theme developed by Edmonds (1981) on the ordered water channels. Based on X-ray diffraction, electron microscopy and biochemical studies, the Ach receptor appears to have four subunits (2:1:1:1) that span the membrane and enclose a channel (Section 7.4). In Edmonds's model five or six subunits are needed to support the pentagon and hexagon water channels, and a channel diameter of ~1 nm agrees well with that of the dodecahedral channel. Furthermore, 0.6 nm repetition length of dodecahedral channels compares well with

---

"support" hydrogen bonds are shown as a dotted line radiating from its center. (b) A sketch of a section through a possible channel support structure consisting of 5 helical rods that enclose the channel and span the membrane. Reproduced with permission from Edmonds (1981).

0.52 nm or 0.63 nm repetition lengths of the presumably helical subunits along the axis inferred from X-ray diffraction studies on the Ach receptors (Ross et al., 1977). (With respect to structured water, it is of interest that X-ray diffraction studies of crystals of a deoxyribonucleoside complexed with the mutagen proflavin showed large-scale water molecules linked together in a regular framework of edge-linked hydrogen-bonded pentagons [Neidle et al., 1980].) The model proposed by Edmonds merits experimental verification and further studies.

Future studies are also likely to be aimed at the isolation and characterization of membrane proteins involved in ion transport to understand the mechanism(s) of transport of various ions across membranes. Recently, Weigele and Barchi (1982) purified sodium channel protein from the rat sarcolemma by using ^{3}H-STX. When this protein was incorporated into lipid vesicles, the resulting vesicles showed increased ^{22}Na$^+$ influx in the presence of BTX, which was specifically blocked by STX. This protein apparently is the essential component of the sodium channel. When it is incorporated into lipid vesicles, the freeze-fractured faces showed 10 nm intramembranous particles. In the absence of the protein, the fractured faces appeared smooth, thereby indicating that the intramembranous particles could be assemblies of the sodium channel protein (see Chapter 3 for a discussion of the nature of the intramembranous particles). Isolated from mammalian brain, the sodium channel protein has been reported to consist of two peptides of MW 270,000 and 37,000, both of which bind the sea anemone toxin (see Hider, 1981).

There are two classes of voltage-sensitive sodium channels: (1) those responsible for the generation of cation potential in nerve and adult muscle, which are specifically inhibited by a nanomolar concentration of TTX and STX, and (2) those found in cultured rat skeletal muscle and denervated mammalian skeletal muscle, which are 100-fold less sensitive to inhibition by TTX and STX and are referred to as TTX-insensitive sodium channels. Lawrence and Catterall (1981) compared the properties of these channels in respect of the action of a number of neurotoxins, for instance, TTX, STX, BTX, veratridine, aconitine, scorpion toxin, and sea anemone toxin II. They reported that these neurotoxins act in a qualitatively similar manner on both classes of the sodium channels, thereby indicating that there is marked structural and functional homology between them. Lawrence and Catterall proposed that the regulation of these channels may be through a post-translational modification (Lawrence and Catterall, 1981). (This homology between the sodium-sensitive and sodium-insensitive channels is reminiscent of the similarity between junctional and extrajunctional Ach receptors [Section 7.1] which are very similar structurally but can be distinguished by isoelectric point and certain antisera [Nathanson and Hall, 1979; reviewed by Malhotra, 1981].) Monoclonal antibody against the saxitoxin binding component from the electric organ of the eel *Electrophorus electricus* specifically binds to a polypeptide of MW 250,000 in the electroplax membrane (Moore et al., 1982). It should be of interest to extend such investigations with monoclonal antibodies to determine the presence, if any, of low-molecular-weight components in the sodium channel, to define the role of the channel components and to determine a homology between sodium channels in various tissues.

The availability of genetic mutations that effect specific membrane functions makes it feasible to investigate the structure and function of the specific components associated with particular functions. For example, temperature-sensitive paralytic mutants of *Drosophila melanogaster* are known that behave normally at a permissive temperature (23–25°C) and become paralyzed at a restrictive temperature (29–37°C). In the *nap* and *para* mutants, the axonal conduction, effected as in the larvae, is abnormally sensitive to TTX. However, these mutants differ somewhat from each other since at room temperature, *para,* unlike *nap,* do not differ from the wild type in TTX sensitivity. It is therefore conjectured that the sodium channels are comprised of at least two different protein subunits—one altered by *nap* and the other by *para* (Wu and Ganetsky, 1980).

Ulbricht (1981), who reviewed the action of drugs on the plasma membrane at the node of *Ranvier,* states that the basic design of sodium channels once introduced has not changed during evolution. However, there are subtle differences in the activation and inactivation (gating time) kinetics in different membranes. Ionic selectivity has been speculated to be a function of the channel macromolecules per se, whereas variations in the gating time constants could be influenced by the fluidity and surface charge of the lipid matrix, whose composition seems to vary among membranes. Thus, TTX, which blocks sodium channels when applied externally, may not be influenced by the lipid matrix, whereas veratridine, whose effect is dependent on temperature, possibly a phase transition, may be.

A behavioral mutant of *Paramecium tetraurelia* manifests an abnormal response when placed in solutions containing Ba^{2+} (Forte et al., 1981). The phospholipid composition of the ciliary membrane of this mutant is markedly different from that of the wild type in its relative amounts of sphingolipid and phosphonolipid. The conductance of both voltage-sensitive Ca^{2+} channels and voltage-sensitive K^+ channels is greatly reduced in the mutant. The behavior, electrophysiological properties, and lipid composition of the mutant plasma membrane become indistinguishable from the wild type when the mutant is grown in the sterol-supplemented medium. These observations imply a significant influence of the membrane lipids on the molecules involved in the generation of action potentials. Such an influence could be exerted by acting on the protein of the ion channel or by regulating the fluidity of the membrane proteins or by contributing to the surface charge or ion binding capacity of the ciliary membrane.

Future studies might also lead us to the discovery of ion-conducting channel proteins in natural membranes comparable to the ion-conducting polypeptides such as nonactin, valinomycin, alamethicin, gramicidin A, and nystatin, which conduct ions selectively when incorporated into lipid bilayers (see reviews by Finkelstein, 1972; Malhotra, 1980). Nonactin and valinomycin serve as carriers and selectively conduct K^+ (Kilbourn et al., 1967), whereas gramicidin A forms a channel selective for monovalent cations. This channel is formed by a dimer of gramicidin A (MW 1879), and its dimensions are 0.5 nm in diameter and 3.2 nm in length. When complexed with Cs^+, the channel widens (0.68 nm in diameter) and shortens (2.6 nm; Koeppe et al., 1978). Nystatin forms pores in the lipid membrane and requires sterol: These pores conduct anions. The most-studied natural oligomeric transport proteins are the (Na^+, K^+)

ATPase, and Ca^{2+}-ATPase, but the true aggregation state of neither of these two proteins is established (see Kyte, 1981). It is thought that (Na^+, K^+) ATPase exists in a dimeric form, whereas the Ca^{2+}-ATPase is in a tetrameric form.

10.2. CALCIUM CHANNELS

Calcium mediates a large variety of cellular functions (see Hagiwara and Byerly, 1981). Some examples are secretion of neurotransmitter (Katz and Miledi, 1967) and hormones (Brandt et al., 1976; Kaczmarek et al., 1980), egg activation (Steinhardt and Epel, 1974), ciliary beat (Dunlap, 1977), and contraction of vertebrate smooth muscle and heart muscle and invertebrate muscle (reviewed by Hagiwara and Byerly, 1981). Axons have a Ca spike during development: Neurites of *Xenopus* neurons have been reported to manifest Ca^{2+}-dependent action potentials in less than 10 hour old cultures. Na spikes appeared after 12 hours in culture, and no Ca^{2+}-dependent action potential could then be observed. In contrast to the neurites, the cell bodies manifested both Na^+- as well as Ca^{2+}-dependent action potentials after the former appeared following 15 hours in culture (Willard, 1980). In view of this, Ca channels are being extensively investigated with a view to understanding the molecular structures and their regulatory control (Fig. 10.2).

Hagiwara and Byerly (1981) have reviewed the current status of the voltage-dependent Ca channel, which is referred to as the channel in the membrane whose permeability to Ca^{2+} increases with a depolarization of the membrane. This channel is distinct from the channels that are not specific for divalent cations, for instance, the channel in the acetylcholine receptor.

As compared with the Na channel, which is almost identical in squid axon, frog nerve, and tunicate egg, Ca channels appear to differ in different tissues, and there may be two Ca channels even in the same tissue (star fish egg) in respect of voltage dependence and kinetics. It is therefore conceivable that the Ca channel is not a single molecular entity (Hagiwara and Byerly, 1981).

Ca channels can be distinguished from other channels, such as Na

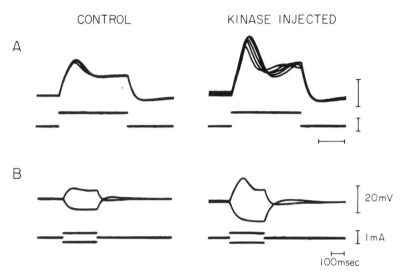

Figure 10.2 The effect of intracellulary injected catalytic subunit of cyclic AMP-dependent protein kinase on the action potentials of bag cell neurons of *Aplysia*. (A) Control shows superimposed oscilloscope tracings of the voltage response (upper traces) to five depolarizing current pulses (lower traces) at a frequency of 0.83 Hz. The right-hand set of traces shows the enhancement of these action potentials that occurs after injection of kinase. (B) This shows the increase in input resistance that may follow injection of the catalytic subunit into bag cell neurons. Control shows the voltage response of a bag cell neuron to constant depolarizing and hyperpolarizing pulses. After injection (right), the voltage responses to the current pulses of the same intensity are increased, the depolarizing current pulse now evoking a graded action potential. Provided by L. K. Kaczmarek.

channels, by a number of electrophysiological parameters, in particular, by the persistence of Ca spikes when Ca^{2+} in the external medium is replaced by Ba^{2+} or Sr^{2+}.

Ca currents are blocked by intracellular Ca^{2+}. For example, in barnacle muscle and snail neurons 5×10^{-7} M intracellular Ca^{2+} greatly reduces the Ca current. However, in egg cells the inactivation of divalent current does not seem to be consistent with this general blockage by intracellular Ca^{2+}, and it is therefore conceivable these channels in egg cell membrane contain a different mechanism for inactivation of Ca currents (Hagiwara and Byerly, 1981).

The regulatory control of Ca currents could lie in the external chemical messengers: For example, extracellular application of nor-adrenaline and isoprenaline increases the magnitude of Ca currents in cardiac muscle. This increase appears to be mediated by intracellular levels of cAMP as the adenylate cyclase is activated by catecholamines (Reuter, 1974). In certain neurons of *Aplysia,* serotonin increases the voltage-dependent Ca current, which action has been reported to be mediated by an increase in the intracellular level of cAMP (Klein and Kandel, 1978). Serotonin inhibits the Ca-dependent secretion of egg-laying hormone from the bag cells of *Aplysia* (Jennings et al., 1981). This action of serotonin affects the Ca channel indirectly. Tetraethylammonium (TEA), which is a blocker of the K channel, has been found to overcome serotonin's effect.

The Na channel is generally considered to be a unique structure whose basic kinetics and selectivity remain unchanged. In contrast, Ca currents vary a great deal in different cells, and it is thought that there may be various types of Ca channels (Hagiwara and Byerly, 1981).

The use of the term *Ca channel* does not necessarily indicate that a channel in nature has been demonstrated. It has been referred to as the voltage-dependent permeation mechanism for Ca^{2+} (Hagiwara and Byerly, 1981).

In addition to the voltage-dependent Ca channel, there is active transport of Ca^{2+} across the plasma membrane (see Barritt, 1981), passive diffusion of Ca^{2+}, and Ca currents that pass through channels that are not specific for divalent cations, for example, the channel opened by acetylcholine (Hagiwara and Byerly, 1981). Some of these aspects of Ca transport are discussed by Barritt (1981).

Based upon estimates of the number of intramembranous particles ($\sim 9.2 \pm 2$ nm) in the freeze-fractured presynaptic membrane of the giant synapse of the squid *Loligo* and the conductance per particle (0.21 ps for a 300 nA current) calculated from presynaptic depolarization, it is feasible that each intramembranous particle represents a calcium channel (Pumplin et al., 1981).

11

Signal Carriers Across the Plasma Membrane

11.1. INTRODUCTION

There are varied manifestations of signal transduction in cells, though a complete understanding of the associated molecular events does not yet exist. A few examples are the hormonal activation of adenylate cyclase (Section 11.3), Ca-dependent release of neuro-transmitter by exocytosis, the opening of cationic channels in the Ach receptors by Ach (Chapter 7), receptor-mediated endocytosis (Chapter 14), chemotatic response of slime mold amoebae to cAMP (Section 11.4), and photoactivation of rhodopsin in the retina (see Saibil, 1982).

11.2. SIGNAL TRANSDUCTION IN RETINAL ROD
OUTER SEGMENTS

Recent studies indicate that cGMP serves a major role in signal transduction from rhodopsin molecules in the membranes of the disks in the retinal rod segments to the sodium channels in the plasma membrane (Fig. 11.1). The disks and the plasma membrane are not physically continuous, and therefore a signal travels from the disks to the plasma membrane. Stryer et al. (1981) have reported the involvement of a small polypeptide (transducin) as the signal carrier. The primary event in the sensory transduction is the photoisomerization of the 11-cis retinal of rhodopsin to the all-transconfiguration. This conformational transition of the rhodopsin results in the transient closure of many sodium channels in the plasma membrane. The hyperpolarization of the plasma membrane thus produced travels to the synapse at the other end of the rod and communicates to other cells in the retina. The response in the plasma membrane, that is, the closure of sodium channels, is highly amplified, as a single photoisomerization of rhodopsin can block the entry of 10^6 sodium ions into the outer segment. It was thought that Ca^{2+}, released from the disks upon the action of light, serves to block sodium channels. However, the evidence that Ca^{2+} serves this role is not conclusive (see Saibil, 1982), and Stryer et al. (1981) provide evidence in favor of cGMP being the transmitter that plays a role in the closure of sodium channels. Favorable evidence comes from the presence of light-dependent cGMP phosphodiesterase in the rod outer segment. The level of cGMP in the rod outer segment decreases upon illumination, and when injected into the cells, it depolarizes the outer membrane of the rod outer segments within milliseconds. In an in vitro system a single photolyzed rhodopsin molecule leads to the hydrolysis of 4×10^6 cGMP per second. Stryer et al. (1981) also determined, by using a nonhydrolyzable analog of GTP, that is, guanosine $5'$-β, γ-imidol triphosphate (GNP), that a single photolyzed molecule in rhodopsin could trigger the uptake of 500 molecules of GNP.

The exchange of tightly bound GDP (guanosine disphosphate) for GTP (guanosine triphosphate) is stimulated by light. In order to

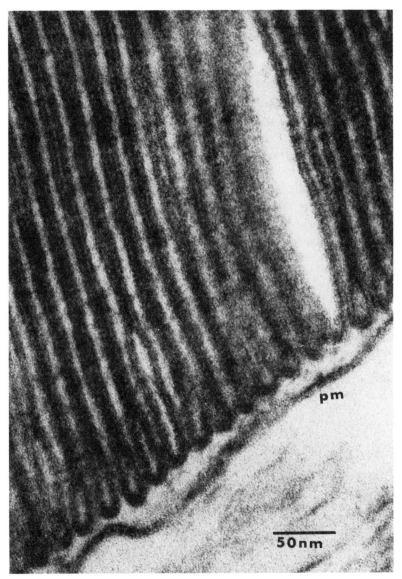

Figure 11.1 Electron micrograph of rabbit retina prepared by freeze-substitution (Van Harreveld et al., 1965) showing a stack of membrane disks in the outer segment of the rod. The rhodopsin molecules are located in the membranes of the disk. Pm = plasma membrane.

account for the finding that there is a highly amplified GTP–GDP exchange, Stryer et al. (1981) searched for the guanyl-nucleotide binding protein, and called this protein transducin (T) since the role of the T-GTP complex is to activate phosphodiesterase. They proposed the following scheme for the mechanism: R* → T-GTP → PDE*, where R* is the photolyzed rhodopsin, T-GTP is the transducin GTP complex, and PDE* the active form phosphodiesterase (Fig. 11.2).

Transducin was extracted from the rod outer segment and SDS-PAGE indicated that it consists of three polypeptides, Tα (MW 39,000), Tβ (MW 36,000) and Tγ (MW 10,000). Purified transducin was incorporated (by incubating it in the dark) into membranes reconstituted from phosphatidylcholine and purified rhodopsin. This

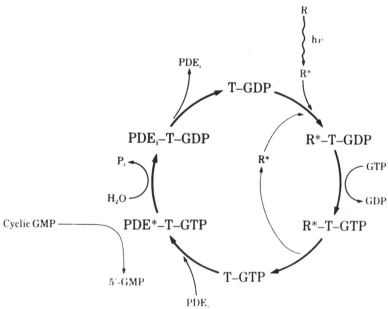

Figure 11.2 Proposed scheme for the flow of information in the light-activated cyclic nucleotide cascade of vision. PDE, phosphodiesterase; T, transducin; R*, photolyzed rhodopsin. Reproduced with permission from Fung et al. (1981).

reconstituted system was functional in the amplified uptake of GNP on illumination. A single photolyzed rhodopsin molecule triggered the uptake of GNP by 70 molecules of transducin bound to the reconstituted membrane. It is therefore apparent that the GTP–GDP exchange of transducin does not involve phosphodiesterase. The α subunit of the transducin contains the binding site for GTP: When the GNP-transducin complex from the reconstituted system was separated into two fractions by high pressure liquid chromatography, one contained GNP and Tα and the other Tβ and Tγ.

Phosphodiesterase is a peripheral membrane protein made up of 88,000, 84,000, and 13,000 MW subunits (see Stryer et al., 1981). Phosphodiesterase bound to the membranes is activated by the addition of T-GNP, which indicates that Tα-GTP is the information-carrying intermediate in the sensory transduction in the rod outer segment. Phosphodiesterase is activated by trypsin, which apparently does so by removing an inhibitory constraint. The 13,000 MW subunit of phosphodiesterase is destroyed by trypsin, which leaves the other two subunits intact, and therefore the 13,000 MW subunit may serve to inhibit the phosphodiesterase. GTP-transducin may activate the enzyme by altering the interaction of the 13,000 MW subunit with the other two subunits. A single photolyzed molecule of rhodopsin activates several hundred phosphodiesterase molecules in two stages. First, T-GTP is formed in an exchange reaction catalyzed by photolyzed rhodopsin; second, T-GTP activates phosphodiesterase by relieving an inhibitor constraint. These two steps can be separated in vitro to show that photolyzed rhodopsin passes information to T-GTP and then to the phosphodiesterase. The gain of 500 in the first step suggests that the transducin is the first amplified information-carrying intermediate in the cyclic nucleotide cascade of vision. The major features of the proposed reaction scheme are given in Figure 11.2. Briefly, in the dark, transducin in the T-GDP form does not activate the phosphodiesterase. Photolyzed rhodopsin (R*) encounters T-GDP on the membrane and forms R*-T-GDP complex in which GTP is exchanged for GDP. T-GTP dissociates from the complex, and R*, thus released, binds another T-GDP. T-GTP activates phosphodiesterase, and the activated phosphodiesterase hydrolyzes cGMP. The GTPase activity of transducin converts T-GTP to T-GDP. The T-GDP and phosphodiesterase are now left in the inhibited form.

The proposed role of transducin in the retinal rod outer segments is comparable to the role of G protein in the activation of the adenylate cyclase system described below. G protein and transducin may thus turn out to be members of a class of proteins that are involved in signal amplification (Stryer et al., 1981).

Besides proteins, phospholipids may serve essential roles in signal transduction across membranes. But the specific molecular details of their involvement do not seem to have been discovered.

Adenylate cyclase is apparently activated in reconstituted systems in the presence of lysophosphatidylcholine and sphingomyelin (Hebdon, 1982). Axelrod (1982) has reviewed evidence for the intramembrane transmethylation of phospholipid as an essential step in signal transduction. Conversion of phosphatidylethanolamine to phosphatidylcholine by transmethylation is thought to be correlated with transductive events in a number of systems, such as the activation of adenylate cyclase complex, the unmasking the cryptic β-adrenergic receptors, and lymphocyte mitogenesis. One of the effects of methylation could be an increase in membrane fluidity. The role of fluidity of the membranes in triggering events in the immune system has been discussed by Metzger and McConnell (1982). Michell (1982) has reviewed the evidence for the involvement of phosphatidylinositol in signal transduction in a variety of systems. It has been proposed that receptor-stimulated turnover of phosphatidylinositol is linked to mechanisms by which cytoplasmic Ca^{2+} concentrations are elevated in stimulated cells, as exemplified by α-adrenergic and muscarinic-cholinergic tissues and antigen and other secretory stimuli in mast cells. Some have speculated that phospholipid methylation is linked to the phosphatidylinositol cycle (Dennis, 1982).

In general, the receptors for various signaling molecules (messengers) such as hormones, neurotransmitters, and Ca^{2+} are membrane proteins. However, gangliosides serve as receptors for choleratoxin (Cuatrecasas, 1973). Whether phospholipid–calcium interactions take place as a mechanism of signaling is not known. The known mechanism of signaling by calcium involves its binding to calcium-modulated proteins in the cytoplasm, for example, calmodulin

(Cheung, 1982a; Kretsinger, 1981). (The transport of calcium through Ca channels is summarized in Section 10.2.)

11.3. THE ADENYLATE CYCLASE-cAMP SYSTEM

A major breakthrough in cell biology emerged as a consequence of studies initiated by Sutherland in the 1950s. Sutherland studied the mechanism of action of hormones, epinephrine and glucagon, on the production of glucose by glycogen breakdown in the liver (1972). An account of the emergence of cAMP as a second messenger in the action of many hormones on the target cells is given by Stryer (1981). It has become apparent that hormonal stimulation of the cellular response through a small molecule such as cAMP results in the amplication of the enzymes that are sequentially regulated, as exemplified by the action of epinephrine or glucagon on liver cells. cAMP is produced from ATP by the action of the plasma-membrane-bound adenylate cyclase. The latter is activated when the primary messenger (hormone) binds to the specific receptor on the plasma membrane. cAMP is hydrolyzed to AMP by specific cAMP phosphodiesterase. Caffeine and theophylline are inhibitors of phosphodiesterase.

Besides glycogen production, cAMP is now known to function in a very wide range of cellular processes, for instance, gene expression in *E. coli,* fat lipolysis in mammalian cells, catecholamine activity (see Greaves, 1977), increase of the secretion of hydrochloric acid by the gastric mucosa, and decrease of blood platelet aggregation (see Stryer, 1981). A review of the role of cAMP in cell growth and differentiation has been published by Friedman (1976). Certain behavioral mutants of *Drosophila melanogaster,* called dunce mutants, have elevated cAMP levels thought to result from a defect in one of the cAMP phosphodiesterase enzymes (Davis and Kiger, 1981): cAMP has been implicated in the assembly of microtubules and microfilaments (Seite et al., 1977; Willingham and Pastan, 1975). cAMP is a primary chemotactic signal in the aggregation and differentiation of the cellular slime mold *Dictyostelium discoideum* (Section 11.4).

The regulatory mechanisms that control the activity of adenylate cyclase are being studied. Adenylate cyclase, which is not directly linked to the hormonal receptor, is activated by a nucleotide regulatory protein, G protein. When G protein is bound to GTP, it activates cyclase; when G-GTP is converted to G-GDP, it is in its inactive form. Thus, GTP–GDP exchange regulates the adenylate cyclase activity (see Stryer, 1981). The physical linkages between the receptor, catalytic unit of the adenylate cyclase, and the G protein are not yet clear. For example, the beta adrenergic receptor and G protein have been obtained in separate soluble fractions, and their recombination was produced in the presence of the hormone (Citri and Schramm, 1980). Nielsen et al. (1981) have discussed the fundamental question regarding the native state of the hormonal receptor, the nucleotide regulatory component (G protein), and the catalytic unit of the adenylate cyclase that converts ATP into cAMP. Though they considered various uncoupled states of the three components in the ground state, their data on turkey erythrocytes favored the precoupled assembly of the three components prior to activation by the hormone.

Rodbell (1980) has reported that the stimulation and inhibition of the adenylate cyclase activity by GTP resides in two functionally distinct regulatory units, and each unit is thought to be linked to a separate class of receptors for hormones and transmitters.

Jacobs and Cuatrecasas (1977) have proposed that the adenylate cyclase is not permanently coupled to the hormone receptor (the mobile receptor hypothesis). According to this hypothesis, the receptors and the enzyme diffuse independently within the plane of the plasma membrane and may reversibly associate with each other. Such a hypothesis is thought to account for the regulation of the single enzyme adenylate cyclase by a number of different receptors, as in the case, for example, when eight different receptors regulate adenylate cyclase in fat cells of the rat (Jacobs and Cuatrecasas, 1977). The mobile receptor hypothesis received support from the results of Orly and Schramm (1976). They fused turkey erythrocytes, in which the endogeneous beta adrenergic receptor had been inactivated by heat or N-ethylmaleimide, with Friend erythroleukemia cells, which have adenylate cyclase but lack beta adrenergic receptors. The fused cells manifested adenylate cyclase activity, as measured by the formation of cAMP, which was stimulated by catecholamine. This stimulation became demonstrable within a few minutes after fusion and was not affected by cycloheximide or emetine.

These results clearly show that the beta adrenergic receptor is not coupled to the adenylate cyclase (Greaves, 1977).

In the respect of the specific role(s) of phospholipids, reconstitution studies on the adenylate cyclase system indicate the nature of the phospholipid environment (Hebdon et al., 1981). When this protein was solubilized from a rat brain homogenate by using sodium deoxycholate, it had no detectable enzymic activity. The maximal restoration of the activity could be obtained by incorporating L-α-phosphatidylcholine, phosphatidyl N-monomethyl ethanolamine, or sphingomyelin. Cholesterol, when present in concentrations above 33 mol %, inhibited the restoration of activity by L-α-phosphatidylcholine (Hebdon et al., 1981). It is also likely that besides specific membrane components, the adenylate cyclase system is regulated by nonmembranous components. Thus, Sahyoun et al. (1981) reported that the detergent-solubilized (Triton shell) adenylate cyclase from erythrocyte of the rat could be reincorporated into erythrocyte ghosts, as well as into liposomes. They suggested that the adenylate cyclase system possesses two domains, one that interacts with the lipid bilayer and the other with the cytoskeleton system of the erythrocyte ghosts. Hagmann and Fishman (1980) demonstrated that the adenylate cyclase activity in the guinea pig macrophages was modulated by the state of assembly and disassembly of microtubules, possibly through some regulatory site (e.g, GTP binding component). Macrophages treated with colchicine manifested up to a 60-fold increase in cAMP response to hormones.

Besides hormones and transmitters, cholera toxin stimulates the adenylate cyclase activity of mucosa of the small intestines, which in turn raises the cAMP level in the cells. This cAMP then stimulates active transport of ions, which results in a large efflux of Na^+ and water in the gut. Adenylate cyclase stimulation appears to be through blocking the GTPase activity of the G protein (Stryer, 1981). A similar stimulation of the adenylate cyclase activity by Vanadate, which is an inhibitor of (Na K) ATPases, has been reported in the fat cells of the rat (Schwabe et al., 1979). While adenylate cyclase activity is stimulated by the action of many hormones, it is inhibited by certain other receptors, as for example, muscarinic, opiate and α-adrenoceptors (see Houslay, 1983). The stimulatory and inhibitory effects on this activity appear to be exerted through distinct nucleotide regulatory proteins. A variant cell line of S49 lymphoma cells (cyc⁻) which has β-adrenoceptors and adenylate cyclase lacks the stimulatory guanine nucleotide regulatory protein and is not stimulated by β-agonists. However, the same cell line does contain an inhibitory guanine nucleotide site and the cells manifest hormone dependent–adenylate cyclase inhibition (Jakobs et al., 1983).

While adenylate cyclase is generally recognized to be a specific marker for the plasma membrane, biochemical and cytochemical studies (Cheng and Farquhar, 1976a, 1976b; Yunghans and Morré, 1978) indicate that it may also be present in the Golgi elements as well as in the smooth endoplasmic reticulum in the rat liver. The Golgi fraction is particularly enriched in cyclase activity. In the fungus, *Phycomyces blakesleeanus,* mitochondrial and nuclear membranes, besides the plasma membrane, showed positive cytochemical reaction product (Tu and Malhotra, 1973). Adenylate cyclase activity has also been reported from nuclei isolated from human lymphocytes. It is conceivable that adenylate cyclase in different locations in cells may be influenced by different cationic and pharmacological controls (Wedner and Parker, 1977).

cAMP controls its varied cellular functions through activation of protein kinases, which bring about phosphorylation of specific substrates (Greengard, 1978, 1979; Schulman et al., 1980). cAMP-dependent protein kinase is a tetrameric enzyme in which two catalytic subunits are associated with two cAMP binding subunits in inactive form. The tetramer disassociates upon binding cAMP and releases the catalytic subunits, which then can transfer the terminal phosphate of ATP to serine or threonine residues on the substrate proteins. In direct support of such a role for cAMP-dependent protein kinase, Kaczmarek et al. (1980) demonstrated by intracellular injection of the catalytic subunit of the kinase into the neurons of *Aplysia* that the calcium action potentials and the protein phosphorylation were both enhanced. The authors conjectured that the phosphorylation increased the channel open time of the voltage-dependent calcium channel and decreased that of the potassium channel.

One of the most extensively investigated cAMP-dependent phosphorylation is protein I, observed in the nervous system (Greengard, 1978). In respect of tissue specificity of the plasma membrane, Greengard (1979) reported that protein I is present only in the synaptic membrane and in synaptic vesicles and has been detected in the neurons from brains of various mammalian species. It is made up of protein IA and protein IB in molar proportion of 1:2, MW 86,000 and 80,000, respectively. Proteins IA and IB appear in the developing brains when synapse formation occurs. Greengard suggested that protein I plays a role in the physiology of the synapse since it serves as substrate for membrane-bound cAMP-dependent protein kinase and a membrane-bound phosphoprotein phosphatase. The phosphorylation of protein I is also calcium dependent (Kennedy and Greengard, 1981). Further analysis of the phosphorylation of protein IA and IB indicated that there are two fragments, one at MW ~35,000 and the other at MW 10,000, that are phosphorylated. cAMP selectively stimulated phosphorylation of the 10,000 MW fragment, whereas Veratridine-induced calcium influx stimulated the phosphorylation of both the fragments. The 10,000 MW peptide has only one phosphorylation site stimulated by cAMP, whereas calcium stimulates the phosphorylation of the 10,000 MW fragment as well as multiple sites on the 35,000 MW peptide. The calcium-dependent phosphorylation requires an endogeneous activator (MW 18,000), which has been identified as calmodulin (Schulman et al., 1980).

It is of further interest that several central nervous system depressants (chloral hydrate, urethane), when administered to mice, produced a decrease in phosphorylation of protein I, whereas the convulsant drugs (picrotoxin) produced an increase. It is therefore conceivable that the phosphorylation of protein I may serve in the functioning of the synapse (Greengard, 1979).

11.4. cAMP AS A PRIMARY SIGNAL MOLECULE

In addition to its role as a second messenger in controlling the activity of a cascade of enzymes, cAMP also acts as the primary extracellular chemotactic signal in the aggregation of the amoebae of the slime mold *Dictyostelium discoideum*. Before cAMP was known, the chemotactic signal was referred to as the amoebae-attracting substance acrasin (Bonner, 1947). This slime mold has been extensively studied as a model system in morphogenesis and in its vegetative phase occurs as single amoebae that feed on microorganisms by phagocytosis and divide every 3–4 hours when food supply is abundant. When the food supply is exhausted the amoebae enter a differentiation phase in which a stalked multicellular organism develops. At the start of the differentiation, the amoebae aggregate by chemotaxis, the chemotactic signal being propagated by cAMP (Gerisch et al., 1975; Konijn et al., 1968; Shaffer, 1975). The aggregation pulses of cAMP trigger other amoebae nearby to propagate the signal. The initial secretion of cAMP is at a frequency of about one pulse every 10 minutes but it gradually increases to one every 2–3 minutes (see Newell, 1977). Extracellular phosphodiesterase secreted by the amoebae avoids permanent saturation of cAMP binding sites. Though no cAMP receptor molecule has been isolated so far, it appears that there are two classes of binding sites, that is, one, a high-affinity binding site and the other, a low-affinity binding site. A vast number of alterations presumably take place in the plasma membrane on the onset of development to transform the membrane from one serving in phagocytosis in amoebae to one responsive in chemotactic signal, aggregation, and function in a multicellular organism. These alterations are reviewed by Bakke and Lerner (1981). While cAMP serves as a chemotactic molecule for *Dictyostelium discoideum,* other genera of slime molds may use different acrasins (Newell, 1977). For example, in *Polysphondylium pallidum,* cAMP serves in the induction of stalk cell formation from undifferentiated amoebae, but it is not chemotactic for this species (Francis, 1975). In *Polysphondylium violaceum,* a peptide (MW 1500) functions as an acrasin (Wurster et al., 1976).

On the basis of the role of cAMP in the development of the cellular slime mold, Robertson and Grutsch (1981) speculate that cAMP may function in development where morphogenesis is believed to be regulated by extracellular signals. For example, in the chick embryo, cell contact formation and movement have been reported to be controlled by a cAMP signal (Gingle, 1977). When stimulated by incubation in cAMP medium ($\sim$ between 8×10^{-8} M and 6×10^{-6} M), dissociated embryonic chick cells secreted cAMP within as little as 5 seconds after stimulation (Robertson et al., 1978). Further explorations should advance our understanding of the extracellular signals that operate during development.

12

Plasma Membrane–Cytoskeletal Interactions

One of the areas in membrane biology where major advances are likely to take place in the near future is in the role that elements traditionally regarded as cytoskeletal (microfilaments, intermediate filaments, and microtubules) serve in controlling functions associated with the plasma membrane (and intracellular membranes). It is now known that actin (microfilaments) and myosin, tubulin (microtubules), and various other proteins that interact with these elements are common cytoplasmic structures of eukaryotic cells (Gordon et al., 1978; Lazarides, 1980). Craig and Pollard (1982) have reviewed interactions between actin and actin-binding proteins that regulate actin polymerization and formation of actin bundles in nonmuscle cells. Some of these cytoskeletal proteins interact with the membranous components and seemingly regulate their functions. Such interactions between the cytoskeletal elements and membranous components may serve in cell motility, cell shape, pinocytosis, secretion, patch formation, anchorage of membranous proteins, and regulation of enzymatic activity.

Cell growth and movement and cell recognition are all thought to be coordinated by an assembly of interacting macromolecules which involve receptors at the cell surface and submembranous fibrillar structures (Edelman, 1976). The fibrillar structures include microtubules and microfilaments. Microtubules are thought to be involved in the anchorage of transmembrane receptors and the propagation of signals to and from the cell surface, whereas microfilaments are thought to be involved in the coordination of the movement of membrane components. Kupfer et al. (1982) have suggested that cytoplasmic microtobules may also serve as tracks along which membranous vesicles derived from the Golgi apparatus travel toward the plasma membrane. These suggestions are based on studies of cultured fibroblasts that had been experimentally wounded. A reorientation of the microtubule organizing center was examined by using immunofluorescence microscopy in the region of the wound where cell movement was directed toward growth of new plasma membrane. Further, Reaven and Azhar (1981) found that the microtubular protein (colchicine-binding protein) from rat brain, when incubated with various membrane fractions from hepatocytes, showed binding, as judged by a lack of microtubule assembly. Such binding was more pronounced when the membrane fractions were derived from the Golgi complex, the cell surface, and mitochondria. Studies on liposomes generated from isolated phospholipids indicated that negatively charged phospholipids (cardiolipin and phosphatidyl serine) were uniquely active in preventing microtubule assembly. The precise significance of such interaction between the membrane phospholipids and microtubular protein is not known: It has been speculated that this protein may be involved in the delivery of phospholipids to the membranes. Bernhardt and Matus (1982) demonstrated that a high-molecular-weight protein (HMWP) associated with microtubule may be specifically associated with the differentiation of dendrites in the purkenji neurons since this HMWP is absent from the cytoskeleton elements in the axon. We know from immunocytochemistry that this HMWP is associated with the growing tips of the dendrites. From this finding, it is apparent that not only the cytoskeletal proteins themselves but also cellular constituents associated with them are likely to influence the functional roles of cytoskeletal elements. It is thus relevant to remark that multiple tubulin isoelectric forms have been detected in a single type of cultured neuron (Gozes and Sweadner, 1981). These observations have been interpreted as suggestive of the presence of structurally different forms of microtubules in a single cell.

The role of the cytoskeleton in regulating the functional organization of the plasma membrane is also illustrated by the reduction of Na and Ca spikes by the breakdown of microfilaments and microtubules, respectively. Fukuda et al. (1981) demonstrated that cytochalasin B and colchicine added to the culture medium specifically reduced, respectively, Na and Ca spikes in guinea pig dorsal root ganglia neurones. The V max of Na spikes was 34 ± 11 versus 121 ± 25 (V s^{-1}) in the controls, whereas the V max of Ca spikes was 4.1 ± 1.8 versus 8.7 ± 2.8 (V s^{-1}) in the controls. These findings suggest specific interaction between the Na channel molecules and microfilaments and between Ca channel molecules and microtubules. The nature of the particular molecular interactions between these cytoskeletal elements and the ionic channels remains to be understood. In the early stages of development of skeletal muscle cells in cultures, Ach receptors can be mostly solubilized when extracted with detergent Triton X-100. However, with maturation, only a proportion of the receptors can be extracted. Such a change in the sarcolemma coincides with the formation of hot spots (patches) of Ach receptors, and it is thought that the receptors within a patch are bound to the cytoskeletal elements and thus resist extraction (Prives et al., 1982). On similar lines, the restricted mobility of the antigenic molecules and Con A receptors in erythrocytes in the adult is thought to result from the interaction of the plasma membrane with the cytoskeleton (Kehry et al., 1977).

The interaction of the plasma membrane protein with the cytoskeletal components is best understood in the rbc (Lux, 1979; see below) where such interactions are considered to play a role in controlling the shape of the cells as they traverse through the blood circulatory system. It is likely that such interactions may be of more general occurrence in animal cells. In this respect the findings of Mescher et al. (1981) are of particular significance. They have reported that a detergent (Nonidet P-40) insoluble complex from murine tumor and lymphoid cells could be pelleted at 100,000 g. The complex contains five polypeptide chains which make up 20–30% of the membrane protein and remain insoluble even at a detergent to protein ratio of 30 : 1. To minimize proteolysis, protease inhibitors were included in the preparation of the membrane fraction and subsequent solubilization. The five polypeptide chains analyzed further are of MW 70,000, 69,000, 42,000, 38,000, and 36,000. Only the 42,000 MW protein was detected in the soluble as well as insoluble membrane fraction, and it comigrated with the rabbit skeletal muscle actin. None of the other four proteins were identified. These proteins do not appear to be exposed on the cell surface as they are not labeled by lacto peroxidase-catalyzed iodination in the intact cells, whereas they are labeled in the isolated membrane fractions. Because most of the cell surface marker enzyme '5-nucleotidase remains complexed with the detergent-insoluble complex, it is suggested that these proteins are components of the plasma membrane.

The morphology and electrophysiology of the tight junctions in *Necturus* gall bladder were found to undergo alterations after exposure to cytochalasin B and phalloidin (which are microfilament active drugs). These alterations are reversible and are suggestive of the involvement of cytoskeletal elements in the regulation of permeability properties of tight junctions (Bentzel et al., 1980).

By using double indirect immunofluorescence, Ball and Singer (1981) demonstrated that in the normal NRK fibroblasts there is an extensive correspondence in the distribution of microtubules and intermediate filaments (vimentin type). Such a correspondence was disrupted in the fibroblasts infected with a temperature-sensitive mutant (LA23) of Rous sarcoma virus within an hour after the cells were transferred from a nonpermissive temperature (39°C) to a permissive temperature (33°C). The authors suggest that the microtubules and microfilaments are linked to the normal cells and that this linkage is affected in the transformed cells.

In respect of interactions between plasma membrane and cytoskeletal elements, the finding that a glycosphingolipid is associated with a colchicine-sensitive microtubule-like structure is particularly relevant (Sakakibara et al., 1981). Glycolipids are generally considered to be present in the plasma membrane to a much greater extent than in the other cellular membranes (Lackie, 1980); the authors confirmed the presence of glycolipids (galactocerebrosides) in the cell line JTC-12 (from monkey kidney). An affinity-purified antibody against galactocerebroside (by immunofluorescence or by the immunoperoxidase method) stained numerous cytoplasmic fibers. On pretreatment with colchicine (but not cytochalasin B) the cytoplasmic fibers were lost, which indicates that the stained cytoplasmic fibers were microtubules. It is of interest to extend this work to confirm the precise site of interaction between the glycolipid and the microtubular system, as such an interaction between plasma membrane and cytoskeleton might facilitate our understanding of the mechanism of transmission of signals between the cell surface and the cell interior.

Geiger et al. (1981) reported that a peripheral protein (vinculin) probably plays a role in the linkage of microfilaments to the membrane. Vinculin is a protein of MW 130,000 (isolated from chicken gizard) and thought to be present in several different cell types. In the intestinal epithelial cells, the microvilli of the brush border have microfilaments extending through their cores and attached at the tips and along their lengths. These microfilaments from various sets intermingle in the terminal web, which then appears to interact with the plasma membrane. This interaction appears to be in the region of the junctional contacts. Immunofluorescence and immunoelectron microscopy reveal that vinculin is densely labeled in the region of the zonula adherens, in contrast to labeling with other cytoskeletal proteins, for example α-actinin or tropomyosin.

It is conceivable that the association of the contractile proteins with the plasma membrane involves Ca binding sites on the plasma membrane (see de Chastellier and Ryter, 1981). X-ray microanalysis of the plasma membrane of the slime mold *Dictyostelium* during differentiation revealed Ca-dependent deposits in gluataraldehyde-calcium-fixed material. Such deposits were found to decrease in the presence of cytochalasin D, which inhibits actin polymerization.

The cytoskeleton attached to the rbc membrane and consisting of filaments and particles has been visualized by electron microscopy of freeze-dried replicas of lysed ghosts fixed in formaldehyde and freeze-fractured under liquid nitrogen (Nermut, 1981).

Caution is warranted, however, in the interpretation of electron micrographs depicting cytoskeletal elements: Small (1981) demonstrated that fixation in OsO_4 (1%) routinely used for postfixation disrupted the actin filaments in the leading edge of the cultured fibroblasts. The actin filaments in the stress fibers were not affected by OsO_4, but dehydration in acetone or ethanol (with or without postfixation in OsO_4) transformed the networks into a reticulum of kinked fibers. In such preparations the leading edge showed a meshwork resembling the "microtrabecular lattice" reported in critical point-dried cultured cells and thought to represent in vivo cytoplasmic organization (Wolosewick and Porter, 1979). Furthermore, Miller and Lassignal (1981) reported that structures resembling well-known biological macromolecules could be seen in nonbiological samples prepared by rapid freezing on copper block cooled with liquid helium and by drying (etching).

Notwithstanding the above criticism, the interaction between the cytoskeletal elements and the plasma membrane is best understood in rbc (Figs. 12.1, 12.2) where it has been demonstrated that there is

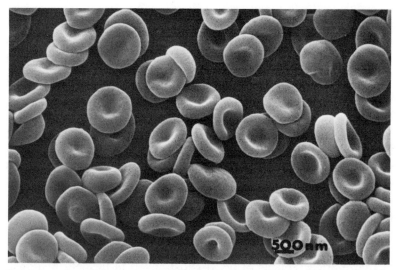

Figure 12.1 Scanning electron micrograph of hamster erythrocytes that were used for analysis of protein profile after preparation of their membrane ghosts (Pimplikar and Malhotra, 1983).

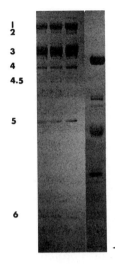

Figure 12.2 SDS-PAGE showing six major bands of protein in the hamster rbc ghosts (three channels on the left) and samples of known MW (on the right): (1) phosphorylase b—MW 90,000, (2) catalase—MW 60,000, (3) ovalbumin—MW 45,000 (diffused), (4) α-chymotrypsinogen—MW 26,000 (arrow at the bottom indicates the dye front). The well-known rbc ghost proteins (see Lux, 1979) are as follows: Band (1) and (2) spectrin; Band (3) protein; (5) actin.

a dynamic submembranous protein network that functions as a membrane skeleton (Gillies, 1982; reviewed by Lux, 1979). The non-ionic detergent extract of rbc consists of an anastomosing network of twisted microfilaments and globular protrusions (see Lux, 1979). Such a cytoskeleton consists of the well-known peripheral protein of the rbc, that is, spectrin (Marchesi and Steers, 1968), actin, ankyrin (Bennett and Stenbuck, 1979), and Band 3 protein, which is a transmembrane membrane. Spectrin is a rod-like molecule (~50 nm × 5 nm) and appears to be involved in the cytoskeleton as a tetramer formed from end-to-end association of heterodimers, the length of a tetramer being 200 nm (Shotton et al., 1979). There are other membrane proteins (labeled as Band 4.1, 4.2, and 4.9) that appear to be involved in the membrane-cytoskeleton complex (Salhany and Gaines, 1981). In rbc, F-actin binding to the cytoskeleton and the plasma membrane is dependent on the presence of Band 4.1. In the absence of this band the inside-out vesicles do not bind actin appreciably, even though spectrin is included in the reconstitution experiments. However, the addition of Band 4.1 restores the F-actin binding, thereby indicating that the binding of the actin to the rbc membrane requires spectrin and Band 4.1 (Cohen and Foley, 1981).

Band 3 protein extends into the cytoplasm, and ~10–15% of this protein is bound by ankyrin, which in turn binds spectrin. G3PD (Glyceraldehyde-3-P dehydrogenase) accounts for ~15% more of the Band 3 protein. Aldolase also binds to the Band 3 at its cytoplasmic NH_2 terminal region (Murthy et al., 1981). Phosphofructokinase is another glycolytic enzyme that appears to bind to the Band 3 protein. The functional significance of the binding of these glycolytic enzymes in the Band 3 protein is not known (Gillies, 1982; Salhany and Gaines, 1981). However, the possibility exists, for example, that such a selective binding may serve to couple substrate transport to synthesis, as Band 3 protein is thought to be involved in the transport of phosphate and phosphate is required for the synthesis of 1,3-diphosphoglycerate by G3PD.

Haemoglobin binds to the cytoplasmic part of Band 3 protein. As haemoglobin competes with G3PD for binding to Band 3 protein, the stilbene disulfonate DIDS (4,4′-bis(isothiocyano)-2,2′-stilbene disulfonate), which is known to bind almost exclusively to Band 3 protein, reduces the affinity of binding for haemoglobin. A parallel inhibition of anion transport and of high-affinity haemoglobin binding can be demonstrated by allowing DIDS to react covalently with the rbc membrane. Also, selective proteolytic digestion shows that the cytoplasmic fragment of the Band 3 protein is needed for the high-affinity binding of haemoglobin. Haemoglobin and Band 3 protein can be cross-linked through S-S bonding between the beta chain sulphydryls of haemoglobin and those of Band 3 protein.

It is of interest that spectrin has been found to be altered in some clinical abnormalities. For example, the spectrin isolated from patients with the disorders described as eliptocytosis (HE) and pyropoikilocytosis (HPP) manifested diminished dimer–tetramer conversions, and the membranes from these erythrocytes were enriched in dimers that may contribute to cytoskeletal instability. Spectrin from HPP patients was different in its peptide map from the normal spectrin in a part of the spectrin known to be involved in tetramer formation (Cohen and Branton, 1981).

Though spectrin has been extensively studied in erythrocytes from higher vertebrates, particularly human beings, it has been detected in the erythrocytes of the most primitive marine animal, *Tere-*

bella, in which the respiratory pigment is enclosed in cells (Pinder et al., 1978). The spectrin was detected by using rabbit antispectrin serum; the cross-reactivity was detected in a single chain of MW 270,000. In mammalian erythrocytes, spectrin is composed of two subunits (MW 240,000 and 220,000) which are similar but not identical (Anderson, 1979). Spectrin or spectrin-like proteins have also been recently detected in several nonerythroid cells that bind actin and calmodulin (see Geiger, 1982). It is therefore likely that a cytoskeleton of the type based on spectrin-actin-ankyrin-membrane protein interactions may emerge as a general feature of the plasma membrane architecture in eukaryote cells. In nonerythroid cells, microtubules and microfilaments provide further complexity in membrane–cytoskeleton interactions that are not yet fully understood.

Apart from the cytoskeleton-plasma membrane interactions, interactions between the plasma membrane and associated extracellular components such as basement membrane, fibronectin, and collagen are likely to play roles in generating or maintaining the functional domains of the plasma membrane. However, our current understanding of these interactions is rather limited (Rajaraman et al., 1978). As an example of how to view the interaction between the plasma membrane and collagen biosynthesis, the studies by Kalcheim et al. (1982) are of particular interest. The researchers found that embryonic rat brain extract, when added to the cultures of myotubes, produced an aggregation of Ach receptors, as well as inducing a 5- to 10-fold increase in the synthesis of procollagen. The procollagen was secreted into the medium and converted to collagen. Furthermore, the presence of ascorbic acid, which stimulates collagen secretion, enhanced the aggregation of the Ach receptors, whereas bacterial collagenase reduced Ach receptor aggregation. In the regeneration of denervated muscle, formation of the postsynaptic structure has been reported to be governed by factors that are external to the nerve (Burden et al., 1979). One of the structures remaining at the synaptic site after removal of the nerve is the basal lamina, the other being Schwann cells. The role of basal lamina and Schwann cells in the formation of postsynaptic structures is not yet certain.

Fibronectins are glycoproteins (MW 200,000–250,000) found attached to cell surfaces as well as in soluble form in the plasma. They have been implicated in cell-to-cell adhesion (Yamada and Olden, 1978), and a small 15 kilodalton fragment of the fibronectin interacts with the rat kidney cells (Pierschbacher et al., 1981).

In addition to the rbc mentioned above, microvilli of the brush border of the intestinal epithelium (Bretscher and Weber, 1980) and formation of the acrosomal process during fertilization in certain invertebrates (Echinoderms; Tilney, 1977) and in the nervous system (Lasek, 1981) are being extensively studied to understand the interactions between the plasma membrane and cytoskeletal elements. However, the nature of the plasma membrane constituents involved in these interactions is not yet certain (see Weatherbee, 1981).

13

The Significance of Plasma Membrane Fluidity

One of the techniques that has been used extensively in determining the extent of fluidity of biological membranes is based on fluorescence photobleaching. Jacobson et al. (1982) have reviewed the application of the fluorescence photobleaching techniques, which enable us to measure the translational mobility of fluorescence-labeled molecules within microscopic specimens. The technique is based on the photobleaching of a discrete region labeled with fluorescent molecules, which destroys their emission. Subsequent fluorescent reappearance in the bleached region is indicative of the diffusion of the unbleached fluorescent molecules from the surrounding regions. The rate of recovery is related, in general, to the diffusion coefficient of the fluorescent molecule.

In artificial lipid bilayers and reconstituted membrane models, the lipids and proteins can apparently undergo a rapid lateral diffusion as the diffusion coefficient ranges from 10^{-8} to 10^{-7} cm^2 S^{-1}. However, in natural systems the cell surface glycoproteins may manifest a very slow diffusion rate or may even be immobile. Lectin-binding glycoproteins on the cell surface of cell lines from kidney have been reported to be immobile (Dragsten et al., 1981, see below). The Ach receptors in the neuromuscular junction are immobile (Section 7.5). The rbc surface glycoproteins diffuse much more slowly than the proteins in the model membranes, if at all. Studies on mature rbc in particular indicate that the control of lateral mobility apparently lies in the interactions between the cell surface proteins and the cytoskeleton. Blebs of membranes of lymphocytes, fibroblasts, and myoblasts which seem to be devoid of cytoskeletal elements showed increased diffusion of proteins (see Jacobsen et al., 1982). (The cytoskeletal elements are best understood in rbc and are elaborated upon in Chapter 12.) A few examples of the postulated role of fluidity in biological processes are described below.

The formation of the Semliki Forest virus and the budding off of the virus particle from the plasma membranes of the host animal cell illustrates the role of fluidity (Simons et al., 1982). This RNA virus has a unit membrane surrounding a nucleocapsid. Extruding through the membrane are spikes. The capsid is made up of 180 molecules of capsid protein (C protein) with one molecule linked to each spike. The latter are made up of one molecule of each of the three glycoproteins E1, E2, and E3. The lipid composition of the viral membrane reflects that of the host plasma membrane. The C protein is synthesized on the free ribosomes and released in the cytoplasm where it associates with the RNA to form the nucleocapsid. E1 is synthesized linked together to an intermediate protein (p62) which consists of E2 and E3. These three proteins are incorporated into the membrane of the endoplasmic reticulum when they are glycoslated. Further glycosylation takes place in the Golgi apparatus from where the spike proteins move and are inserted over the plasma membrane of the host cell (Green et al., 1981). The nucleocapsids move and come to lie against the plasma membrane where their C protein has been thought to encounter a spike, and anchor the capsid to the plasma membrane. Further spikes move laterally and come to link with the C protein until all of the 180 molecules of the C protein have been linked, at which time the virus particle buds off. How the host membrane proteins are excluded from the viral membrane is not yet understood.

There may be specific regions of the plasma membrane from where the virus buds off, a notion based on the demonstration by Rodriguez-Boulan and Sabatini (1978) that various enveloped viruses bud off from different regions of the plasma membrane in polarized epithelial cells. These authors used monolayers of cell lines derived from kidney (dog and bovine) and embryo cells (chicken) and infected them with vesicular stomatis virus (VSV), influenza virus, Sendai virus, or Simian virus 5 (SV5). These viruses contain a small number of membrane glycoproteins, and it is assumed that their biosynthesis and transport to the cell surface takes place by a mechanism essentially similar to that that the uninfected cells employ for their own plasma membrane proteins (Bergmann et al., 1981; Green et al., 1981). Rodriguez-Boulan and Sabatini found that influenza virus, SV5, and Sendai virus bud off exclusively from the apical surface of the cells, whereas VSV buds off from the basolateral plasma membrane. It is of obvious interest to know the specific signals that are used and the mechanism underlying the transport of these glycoproteins to their destination. Rodriguez-Boulan and Sabatini speculate that carbohydrates may serve a role: SV5, influenza virus, and Sendai virus lack sialic acid as do the enzymes in the apical plasma membrane of epithelial cells. VSV contains sialic acid and so does the Na^+, K^+, ATPase in the basolateral plasma membrane.

The role of fluidity is further illustrated in the growth of the plasma membrane at the growth cone in neurite outgrowth. The plasma membrane of the growth cone differs from that of the neuron cell body in lectin binding sites and is relatively devoid of IMPs. Pfenninger (1981) employed a variety of approaches for such an investigation, for instance, pulse-chase experiments with lectins, measurement of the density of IMPs in the plasma membrane of the growing neurites as a function of time and distance from the cell body, and intracellular transport of tritiated phospholipid precursor (^{3}H-glycerol). His data led him to conclude that the growth of the neurite takes place by fusion of IMP-free vesicles with the plasma membrane. These vesicles carry newly synthesized glycoconjugates and are transported from the cell body. Further, the IMPs are inserted in the plasma membrane of the cell body and reach the growth cone by lateral diffusion.

One of the best instances of the role of fluidity and functional asymmetry of the plasma membrane has been reported by Dragsten et al. (1981) who examined the implications of the tight junctions (Section 6.2) as a diffusion barrier in epithelial tissues. They cultured continuous cell lines derived from kidney; these cells form a confluent monolayer. By using fluorescent-labeled lectins (WGA and pea agglutinin) as probes and fluorescence photobleaching recovery estimates, Dragsten et al. (1981) found that the lectins are immobile on the cell surface. Even when the monolayers were dispersed into single cells by chelating divalent cations, the lectins did not diffuse into the photobleached areas. In contrast to the lectins, lipid was found to be mobile and also diffuse through the region of the tight junctions. However, the lateral diffusion of the lipid could take place only if the lipid could undergo flip-flop motion into the inner layer of the plasma membrane. These experiments indicate that the tight junctions present a barrier to the lateral diffusion of lipid in the outer half of the plasma membrane but not to the lipid in its inner half. This barrier to diffusion may result from the fusion of apposed membranes in the region of the tight junction (Kachar and Reese, 1982).

The lateral diffusion of LDL receptors dispersed over the entire plasma membrane and their clustering in the coated pits are necessary for the receptor-mediated endocytosis of LDL particles. Fibroblasts from patients manifesting abnormal cholesterol metabolism show a lack of lateral movement of LDL receptors so that the receptors do not reach the coated pits. The endocytosis of the LDL is thus defective (Brown et al., 1979). (Receptor-mediated endocytosis of LDL particles is a vehicle that provides cholesterol to the mammalian cell.) The ligand-induced lateral aggregation (patch formation) followed by clustering at one pole (cap formation) of surface immunoglobulin of lymphocytes is followed by pinocytosis (Raff and de Petris, 1973).

The directed lateral diffusion of lipids and glycoproteins from the tip of the pseudopod of the spermatozoan of the nematod apparently provides this amoeboid cell with a mechanism for its motility without involvement of actin or myosin (Roberts and Ward, 1982a, b). The hindgut termite flagellate provides another model system to study membrane fluidity. The head of this protozoan rotates continuously relative to its body (0.7 rotations/second) without any apparent structural difference in the membrane between the head and the rest of the cell body (Tamm, 1979). These flagellates carry on their surface, symbiotic rod-shaped bacteria that serve as useful markers for the study of membrane movements (see Tamm, 1979).

A change in the fluidity of the membrane of sea urchin eggs has been reported to accompany activation (Scandella et al., 1982). Spin-labeled eggs (with 5'-doxystearate) showed a decrease in order parameter within 10 minutes after fertilization. The decrease in the order was 2–3%, which is quite dramatic as changes in phase-transition of greater than 1% are apparently rare in biological systems. (Phase transition from gel to liquid crystal in phospholipid involves a change of ~20% [see Scandella et al., 1982].) Scandella and coworkers obtained in vitro preparation from sea urchin eggs consisting of the plasma membrane and cortical granules that are membrane-bound and lie underneath the plasma membrane in unfertilized eggs. Upon fertilization, the cortical granules disappear and are exocytosed, a process which requires an increase in cytoplasmic Ca^{2+} (1–3 μm). The authors report that the in vitro cortices manifest changes in fluidity with the addition of Ca^{2+}. There appear to be two distinct changes in membrane fluidity that follow within 10 minutes of fertilization: a transient increase in cytoplasmic Ca^{2+} concentration that induces a decrease in the fluidity of the plasma membrane and an increase in cytoplasmic pH that causes an increase in the fluidity of internal membrane in the egg.

The Ach receptors undergo rapid lateral diffusion (2.6×10^{-9} cm^2 S^{-1}) in the plasma membrane of cultured embryonic muscle (Poo, 1982). Such a rapid diffusion could provide an efficient mechanism of localization of the receptors when nerve–muscle contact is established during development. Poo (1982) used an electrophysiological approach in which Ach receptors were inactivated by the localized application of α-BGT, and the recovery of Ach sensitivity was monitored in cultures of *Xenopus* embryonic muscle cells. The Ach receptors were found to undergo rapid lateral diffusion (diffusion coefficient 2.6×10^{-9} cm^2 S^{-1} at 22°C). The possibility that inactivated Ach receptors had become functional or that newly synthesized Ach receptors had been incorporated in place of inactivated sites during the time scale of the experiments was ruled out. Also, the Ach sensitive recovery was blocked when cells were preincubated with Con A that binds specifically to α-D-glucose or α-D-mannose residues of carbohydrates. Con A has been demonstrated to bind to Ach receptors, which are then immobilized. This rapid diffusion coefficient of Ach receptors in the embryonic muscle of *Xenopus* is comparable to the diffusion coefficient of rhodopsin molecules in the retinal disks and surface antigens of cultured muscle cells (see Poo, 1982).

It is of further interest that Poo (1982) found no detectable Ach receptors that were immobile. Stya and Axelrod (1981) have also found that the Ach receptors in the clusters of Ach-sensitive spots described as "hot spots" (Fischbach and Cohen, 1973; reviewed in Malhotra, 1981; Tipnis and Malhotra, 1980) become laterally mobile after sodium azide treatment. Moreover, the diffuse Ach receptors become aggregated and form patches on the plasma membrane. The relevance of these findings pertains to previous studies in which a substantial fraction of the Ach receptors were found to be immobile; also, the diffusion of the mobile fraction was 50-fold slower than that detected by Poo (1982). The difference may be in the techniques applied in these studies, as the previous results were based on the photobleaching of fluorescent α-BGT. The possibility then arises that the labeled α-BGT-receptor complex may hinder in some undetermined manner the diffusion of the Ach receptors (see Jacobson et al., 1982; Poo, 1982).

With a such rapid lateral diffusion of Ach receptors as reported by Poo, it is conceivable that an efficient localization of the receptors could be achieved during synaptogenesis. An Ach receptor could diffuse an average distance (S^{-2}) of 150 μm $(S^{-2} = 4Dt$, where D is the diffusion coefficient and t is the time). The mechanism of trapping the Ach receptor molecules in the nerve–muscle contact zone is not known but clustering could be favored by Van der Waals forces (Gingell, 1976). The interaction between nerve and muscle is an important aspect of cell biology now under investigation.

Obviously, there is considerable evidence to support the concept of fluidity in biological membranes. Also, fluidity may play a vital role in serving several roles associated with the membranes. However, a great deal of data are available that indicate that some of the membrane components are relatively immobile. One of the best-studied cases is that of the Ach receptors in the neuromuscular junction in the adult skeletal muscle: The receptors are restricted to the junctional sarcolemma (Chapter 7). Another example is the reported immobility of Na channels (see below).

Stühmer and Almers (1982) made observations on the immobility of Na channels in the sarcolemma of the frog skeletal muscle. They photobleached with UV irradiation (using a fine micropipette) a patch of membrane ~10 μm in diameter and measured Na currents during membrane depolarization. A 30–90 S irradiation reduced Na currents three- to five-fold, and no recovery was detectable for one hour, which indicates that the fresh channels had not moved into the irradiated patch from the neighboring areas of the sarcolemma. Lateral immobility may help maintain uneven distribution of the Na channels over the plasma membrane. Uneven distribution of Na channels is a normal feature of the mammalian myelinated axons where these channels are essentially confined to the node of *Ranvier* and K channels are localized at the internodal plasma membrane (see Chiu and Ritchie, 1980, 1981; Fritz and Brockes, 1981). An estimate of $12,000/\mu m^2$ compared to less than $25/\mu m^2$ has been made for the nodal and internodal plasma membrane distribution of Na channels (Ritchie and Rogart, 1977). The study of the distribution of Na and K channels has been facilitated by the use of specific ligands, saxitoxin and TEA, which block Na and K, respectively.

The various processes of neurons, such as axons, dendrites and synapses maintain their characteristic functional organization presumably by restricting free movement of their components in an otherwise morphologically continuous membrane. Such a functional compartmentalization is comparable to the functional domains in the well-documented epithelia (Section 5.2). There are various other examples of functional domains, some of which are discussed in Chapter 5. The important question remains as to how such functional domains are generated and maintained in a multifunctional membrane continuum. It is conceivable that the cytoskeleton may play some role in such an organization by specific interactions with membrane constituents. But our current understanding is limited to essentially what is known in erythrocytes (Chapter 12).

14

Endocytosis, Exocytosis, and Membrane Fusion

Large amounts of plasma membrane are interiorized through endocytosis either continuously as in pinocytosis or in discrete phagocytic events. Macrophages and L-cell fibroblasts interiorize 186% and 54% of their surface area, respectively, as pinocytic vesicles, each hour without alteration in cell volume or surface area (Silverstein et al., 1977). Such evidence suggests that membrane recycling may take place continuously. The low-density lipoprotein (LDL) receptors that function in cell uptake of LDL have been shown to recycle back to the cell surface (Goldstein et al., 1979). Not all the receptors that are internalized necessarily undergo recycling. In this respect, Das and Fox (1978) demonstrated that the receptors for epidermal growth factor (EGF), which is a mitogen for target cells (e.g., 3T3 cells), are internalized and degraded in lysosomes. The EGF receptor is a single polypeptide chain (MW 190,000). By using radiolabeled photoreactive derivative of EGF, Das and Fox (1978) found that after one hour incubation at 37°C, most of the radioactivity was found in three SDS-PAGE bands (i.e., at MW 62,000, 47,000, and 37,000) that banded with the lysosomal subfraction, thereby indicating that the receptor molecule had been degraded, presumably in the lysosomes. However, the membrane components may recycle by returning to the cell surface.

Diphtheria toxin appears to be taken up by endocytosis (Van Heyningen, 1981). This toxin (MW 60,000) is made up to two polypeptide chains, fragment A (MW 21,000) and fragment B (MW 39,000), linked by a disulphide bond. Fragment A is an enzyme that catalyzes a reaction in which the ADP-ribose of NAD is transferred to an elongating factor 2 (EF2). ADP-ribosylated EF2 is inactive, and thus the cell protein synthesis stops. However, the toxin binds to the cell through fragment B, which then facilitates entry of fragment A into the cell where it undergoes a conformational change. The endocytotic vesicles fuse with the lysosomes, where most of the toxin is perhaps destroyed—yet a few molecules of fragment A escape into the cytoplasm and cause cell death.

Membrane recycling may be a common phenomenon, not only confined to a few glandular cells but common to most eukaryotic cells. Membrane insertion and removal can take place in a nonrandom fashion (Herzog, 1981). By using electron dense tracers (e.g., dextran, cationized ferritin, and horseradish peroxidase), Herzog found that membrane recycling took place by two routes: either through the lysosomes and then the Golgi complex or directly through the Golgi complex. Dextran appeared in the secretory granules 15 minutes after infusion of the tracer through the duct into the acinus lumen. The appearance of dextran in the cisternae of the Golgi complex within 5 minutes indicates that the dextran is transferred presumably via vesicles that are pinched off from the luminal plasma membrane (Herzog and Farquhar, 1977). These endocytotic vesicles fuse with the Golgi membranes to transfer the tracer into the cisternae. These findings provide support to the concept that the membrane reutilization takes place as a normal cellular phenomenon. 5'-nucleotidase, which is a membrane-bound enzyme, has been found to be continuously exchanged between the plasma membrane and the cytoplasm by the recycling of membranes (Widnell et al., 1982). Less than 50% of the total enzyme was detected at the cell surface, whereas the remaining enzyme was present in cytoplasmic membranes.

In relation to the recycling of plasma membrane, it is of interest that Schneider et al. (1981) found that membrane recycling may involve the return of the membrane patches from the lysosomes to the cell surface. It is likely that they pinch off from the lysosome surface in the form of closed vesicles and return to the cell surface by as yet unknown routes. These conclusions are based on studies of the rat embryo fibroblasts cultured in the presence of control rabbit immunoglobulins (C IgG) doubly labeled with ^{3}H-acetylation (A) and then conjugated with fluorescein (F). The IgG molecules are taken up without the mediation of any specific receptors. The fibroblasts take up FAC IgG continuously for at least 72 hours and then return the major part of their intake back to the medium but in the form of low-molecular-weight products. By using a variety of techniques in corroboration, such as gel filtration, immunological analysis, cell fractionation, lysosomal analysis, and incubation of FAC IgG in vitro in the presence of lysosomal extract from hepatocytes, Schneider et al. (1981) concluded that the storage and digestion of FAC IgG occur in lysosomes and that the Fab fragment of three-quarters of the IgG molecules is indigestible and accumulated intracellularly. It would appear that the lysosomes may not be a depository of dead-end material. They are "connected with the extracellular material via mobile membrane patches acting as some sort of endless moving belt, and are capable of transporting materials in both directions" (Schneider et al., 1981, p. 386).

One of the well-known instances of exocytosis is the fusion of synaptic vesicles with the presynaptic membrane (Kelly et al., 1979). Von Wedel et al. (1981) have demonstrated by using immunofluorescent techniques that the plasma membrane of the presynaptic nerve ending and the membrane of the synaptic vesicles in the electric organ of the marine electric ray *Narcine brasiliensis* are different in comparison. Antiserum to purified synaptic vesicles raised in rabbit bound selectively to synaptic vesicles in the nerve–muscle preparations of the frog when the latter was (cryostat) sectioned and treated with fluorescein-labeled goat antirabbit antibodies. In intact nerve–muscle preparations there was no marked labeling, presumably because the antibodies could not permeate the plasma membrane of the nerve terminal and only a few vesicle antigens were accessible at the surface of the nerve terminal in the normal resting state. However, in the condition (induced by use of 1 mM LaCl₃) wherein the synaptic vesicles were stimulated to release their acetylcholine contents, the intact nerve terminals became labeled, which indicated the transfer (by exocytosis, presumably) of synaptic vesicle antigens to the plasma membrane of the nerve terminal.

The factors that govern the route for membrane recycling could include the nature of the contents of the endocytotic vesicles, their net charge, cell type, and the physiological state. To illustrate the significance of the cellular physiological state for membrane recycling, Herzog and Farquhar (1977) reported that in the rat lacrimal or parotid acinar cells, dextran is transferred to the Golgi complex in vivo, but when acini are isolated in vitro and incubated in the presence of dextran, the tracer uptake by the Golgi complex is delayed and uptake by the lysosomes is increased. (These experiments were done under conditions that stimulated the acini to discharge secretory granules; isoproterenol to stimulate secretion in the parotid gland, and carbamylcholine to stimulate the lacrimal gland.)

Membrane vesicles that participate in membrane recycling may belong to the "coated" type when they are associated with clathrin (Section 5.3) or may have a smooth surface when they lack clathrin. Both types of vesicles have been implicated in the transport of receptors and antigens from the cell surface to the lysosomes (see Geisow, 1980).

The role played by coated pits in the internalization of receptor-mediated macromolecular ligands is a case to be considered in some detail. Coated pits can be easily recognized in electron micrographs as indentations in the plasma membrane and coated with a densely staining fuzzy material (Fig. 5.2). Such pits have been demonstrated to contain receptors for many different macromolecules, for example, polypeptide hormones (insulin, epidermal growth factor, nerve growth factor; Schlessinger, 1980), glycoproteins terminating with mannose of N-acetylglucosamine and lysosomal hydrolases (Stahl and Schlesinger, 1980), maternal immunoglobulins (Rodewald,

1973), ferritin and transferrin (Pearse, 1982), and LDL (Goldstein et al., 1979).

It has been suggested that coated pits may serve as molecular filters that selectively allow some plasma membrane proteins to be internalized while others are excluded. In this context, Bretscher et al. (1980) demonstrated that in the fibroblast cell line N1H 3T3 cells, two proteins (i.e., θ and H63 antigens) are excluded from the coated pits. These proteins were detected by labeling with ferritin-conjugated antibodies and found dispersed on the plasma membranes, but coated pits were largely devoid of the ferritin label. In contrast to these two antigens, the transmembrane protein receptors for LDL are well known to be concentrated within the coated pits (Goldstein et al., 1979). These LDL receptors are rapidly internalized during the process of transport of LDL into the cell. Each receptor is internalized every 5–10 minutes at 37°C (Anderson et al., 1981). It is thought that there may be other cell surface receptors that are also concentrated in the region of the coated pits along with the LDL receptors. Cholesterol may also be excluded from internalization. In support of exclusion of cholesterol from the coated pits, Montesano et al. (1979) found that filipin-cholesterol complexes were absent from the pits in the plasma membrane visualized in freeze-fracture replicas, whereas the remaining plasma membrane had distinct protuberances (25–30 nm) characteristic of the sterol-filipin complex. Obviously, the sterol content of the plasma membrane in the region of the pits is much less than in the rest of the plasma membrane (Pearse, 1976). Bretscher et al. (1980) also proposed that coated vesicles may play a role in continuous removal of lipid from the plasma membrane, which could be reinserted along with the incorporation of membrane proteins.

That bulk membrane flow occurs has been demonstrated by Roberts and Ward (1982a). The differentiation of amoeboid spermatozoa of the nematode *Caenorhabditis elegans* is a case for the study of fluidity and bulk membrane flow (Roberts and Ward, 1982a). The sessile spherical spermatid is transformed into a motile spermatozoan that consists of a hemispherical body and a 3–4-μm-long pseudopod. This differentiation, which can be induced in vitro by using monensin (monovalent ion ionophore), involves extensive reorganization in the cytoplasm and fusion of membranous organelles with the plasma membrane. This fusion occurs at the base of the pseudopod. By using fluorescent-labeled lectins, ferritin-labeled lectins, and fluorescent phospholipid analogue, Roberts and Ward (1982a) reported that the pseudopodia are initially labeled with the surface markers but within a few minutes the markers gradually disappear from the pseudopodia. The cell body remains uniformly labeled, thereby indicating that the markers on the plasma membrane of the pseudopod had flowed into the cell body. Roberts and Ward (1982b) further showed that the lectin-labeled receptors are continuously inserted at the tip of the pseudopod and move laterally to its base where they are presumably internalized. (The mechanism of insertion of the lectin receptors is not known.) However, the bulk membrane flow has been suggested as the mechanism that provides for the motility of these spermatozoa: Actin contents of the sperm are <0.02% of the total protein. No myosin or microfilaments were detected (Nelson et al., 1982).

Coated pits and coated vesicles receive their name from their bristle coat appearance, which has been found to contain a characteristic polypeptide called clathrin (MW 180,000; Pearse, 1975, 1976). The amino acid composition of the clathrin from various sources (pig brain, bullock brain, and bullock adrenal medulla) indicates that clathrin is highly conserved in its composition (Pearse, 1976). It is generally believed that the coated pits pinch off and form coated vesicles, thus internalizing the receptor-linked extraneous macromolecules. However, the observations made by Wehland et al. (1981) imply that the entire coated pit may not pinch off from the plasma membrane. They reported that antibodies to anticlathrin, when microinjected into cultured fibroblasts, did not prevent the endocytosis of α_2-macroglobulin, which is a receptor-mediated process and associated with the coated pits. The authors considered that the coated pits are stable and permanent features of the plasma membrane. The internalization of the plasma membrane is believed to be through the formation of an endocytotic vesicle by growth of the membrane either adjacent to or from within the coated pit.

Clathrin is capable of structural transformation. Purified brain clathrin can occur in two forms: the "basket form" characteristically seen in electron micrographs surrounding the coated vesicles and a filamentous form (Schook et al., 1979). These polymeric forms are interconvertible with small changes in pH (7.5–6.5). A similar transformation in vivo might occur and function in the process of endocytosis in association with the coated pits. Clathrin in vitro (clathrin-coated polystyrene particles) bound some of the muscle proteins (F-actin, G-actin, and α-actinin) but not tropomyosin and serum albumin. These findings indicate the selectivity of interactions between clathrin and other proteins. Schook et al. (1979) proposed that coated structures are associated with the cellular actomyosin system, and interactions between the various components of the system may serve in motility, force, or guidance for vectorial translocation of the coated vesicles.

Exocytosis, endocytosis, and membrane recycling involve fusion of membranes, which also takes place in a variety of other cellular phenomenon such as fertilization, cell division, formation of multinucleate myotubes from mononucleate myoblasts, and the budding off of certain virus. The molecular events and the factors that play a role in fusion are not fully understood. Whether a general mechanism is operational in fusion does not seem to be clear (Fig. 14.1).

Membrane fusion has been induced experimentally by various factors, including Sendai virus, temperature (~50°C), lysolecithin (Lucy, 1975), pH (White and Helenius, 1980), calcium (Ingolia and Koshland, 1978), and alamethicin (Lau and Chan, 1975). Polyethylene glycol (PEG) is now routinely used as a fusogen for producing hybridoma cells by fusing myeloma cells and immunized spleenic cells. However, fusion of two membranes will involve close apposition and destabilization of membrane structure. Ca^{2+} may facilitate both of these events in membrane fusion (Papahadjopoulos et al., 1978). Fusion of membranes may be preceded by an increase in membrane fluidity (Poste and Allison, 1973; Prives and Shinitzky, 1977). Destabilization can arise if the membrane lipids undergo phase transition from bilayer to a nonbilayer micellar form in the fusion zone (Lucy, 1975). That a hexagonal H11 form, consisting of inverted micelles, exists in experimental systems has led to the interpretation that a similar transition takes place during fusion of biological membranes. Such a transition requires Ca^{2+} (Cullis and

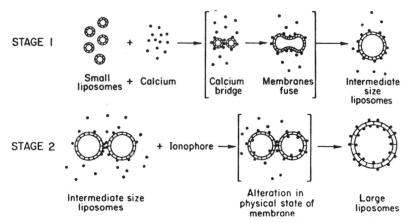

Figure 14.1   Model for calcium-induced fusion. Fusion occurs in two stages. In the first stage, calcium ion binds to phospholipids on the outside of small (approximately 60 nm diameter) liposomes and catalyzes fusion events that produce intermediate-sized liposomes (approximately 120 nm diameter). Calcium may be exerting its effects by altering the membrane properties so that fusion is facilitated and by bridging the liposomes so that the liposomes are held in close proximity. The second stage of fusion occurs when calcium ion can traverse the membrane, such as in the presence of ionophores. Calcium may catalyze this stage of fusion by binding to the phospholipids on the inside of intermediate-sized vesicles, thereby altering the membrane properties once again, to allow formation of large liposomes (approximately 170 nm diameter). Reproduced with permission from Ingolia and Koshland (1978).

Hope, 1978). Whether such a phase transition takes place in fusion of natural membranes remains to be ascertained.

A phase transition of the type mentioned above would be facilitated by lateral segregation of membrane proteins from the fusion zone to create a protein-free lipid bilayer (Ahkong et al., 1975; Cullis and Hope, 1978). Through electron microscopy of freeze-fracture replicas of pancreatic β-cells, it has been demonstrated that the fusion zone during exocytosis of secretory granules is devoid of IMPs (Orci et al., 1977). In support of this, Lawson et al. (1977) applied ferritin-conjugated lectins (PHA) and anti-immunoglobulin antibodies to study membrane fusion in the rat peritoneal mast cells, which discharge their secretory granules by exocytosis. They observed that the label was absent from the region of the plasma membrane that interacts with the membrane of the secretory granules, either prior to fusion or after fusion with the granule membrane. These observations indicate that the protein is lacking in the zone of fusion when the membranes fuse. However, these observations have been made on tissues that were chemically fixed and treated with glycerol prior to freezing. The possibility then arises that the IMPs may have been displaced during preparations. These remarks are supported by the observations of Ornberg and Reese (1981), who found IMPs in the fusion zone by using rapid freezing without any prior chemical treatment. Their findings are based on the exocytosis of secretion in *Limulus* amoebocytes induced by endotoxin. Proteins may not necessarily be cleared from the fusion zone. Also, clustering of intramembranous proteins seen in freeze-fracture replicas might serve to facilitate fusion by interdigitation of protein molecules on apposed membranes, which could then spread to include lipid bilayer regions (Poste and Allison, 1973). However, the role played by specific lipids and proteins in recognition of fusion sites or in the process of fusion in vivo is not yet certain. The plasma membrane may undergo specific changes in its composition during fusion. Howard et al. (1982) demonstrated that two proteins of MW 45,000 and 50,000 preferentially appear in the plasma membrane of macrophages, while one protein, MW 140,000, disappears after antibody-dependent phagocytosis of lipid-hapten containing lipid vesicles. Cytoskeletal filamentous elements may also be involved in endocytosis or exocytosis (Boyles and Bainton, 1981; Ornberg and Reese, 1981). However, apart from the involvement of Ca^{2+} in fusion, details of the mechanism are not yet clear enough to generalize events that are associated with fusion.

Salisbury et al. (1981) demonstrated the involvement of calmodulin in receptor-mediated endocytosis of Con A receptors in a human lymphoblastoid cell line. By using fluorescein Con A and rhodamine-anticalmodulin antibodies the authors observed a diffused intracellular distribution of calmodulin in cells fixed immediately after Con A challenge. However, during capping, calmodulin appeared concentrated at sites of endocytosis. This redistribution was dependent on Ca^{2+} because incubation of cells in a Ca^{2+}-free medium prevented the redistribution of calmodulin, though capping of Con A preceded normally. Capping was blocked by cytochalasin though the initial redistribution of calmodulin was not affected. These observations suggest that calmodulin may be involved in the movement of clathrin to the cell surface at the site of the ligand receptor complex.

Perhaps the best available evidence for the involvement of calmodulin in the fusion of biological membranes has come from studies on the exocytosis of cortical granules in the sea urchin egg. At fertilization the cortical granules fuse with the plasma membrane and discharge their contents into the extracellular space. This process of exocytosis involves intracellular Ca^{2+}. Steinhardt and Alderton (1982) demonstrated by using antibodies to calmodulin and immunofluorescence microscopy that calmodulin is associated with the inner surface of the plasma membrane as well as with the cortical granules, perhaps where the granules attach to the plasma membrane. They also demonstrated that the antibodies to calmodulin inhibit Ca^{2+} sensitivity of exocytosis. (A change in the fluidity of the plasma membrane that takes place upon fertilization of the sea urchin egg is mentioned in Chapter 13.)

Membrane fusion is discussed in detail by Gratzl et al. (1980) and Poste and Allison (1973), and calcium control of exocytosis and endocytosis is elaborated by Baker and Knight (1981).

15

Biosynthesis

15.1. INTRODUCTION

All cellular membranes are assembled by inserting newly synthesized components into preexisting membranes, there being no evidence so far in favor of a de novo synthesis of membranes in vivo.
The functional domains could turn over as individual units or their
lipid and protein components could turn over at rates characteristic
of each type of lipid and protein. In the latter case, there would be
no harmony in the turnover rates of various membrane components
(reviewed by Malhotra, 1976). Furthermore, within a single class of
lipid, there may be more than one kinetically distinguishable population of molecules, each with its own turnover rate (Finean et al.,
1978). The turnover rates are complicated by the recycling of membrane constituents and by the fact that the turnover rates are often
derived from membranes of functionally heterogeneous population
of cells (Tweto and Doyle, 1977).

15.2. LIPID BIOSYNTHESIS

Plasma membrane lipids are synthesized in the endoplasmic reticulum. (Mitochondria and chloroplasts may have the necessary enzymes for the synthesis of some of their own lipids [Finean et al., 1978].) The plasma membrane lipids are glycosylated in the Golgi apparatus and possibly in the plasma membrane itself. From the Golgi apparatus, they are transported presumably via vesicles which fuse with the plasma membrane.

One of the questions regarding the biosynthesis of membrane lipids is the origin and maintenance of their asymmetric distribution. We know from observations of *Mycoplasma* (Rottem, 1981) that cholesterol is translocated from one half of the bilayer to the other. Phospholipids may also undergo similar translocation. But, the important question that remains to be answered is, what is the mechanism of preferential translocation of phospholipids from one half of the bilayer to the other half? Such a mechanism might be operational if the biosynthesis were to occur preferentially in the cytoplasmic half with a rapid translocation to the outer half. The findings of Marinetti and Cattieu (1982) on the rbc are relevant to the maintenance of the asymmetric distribution of phospholipids. Using vectorial probes and radioactive fatty acids, they examined the turnover of phosphatidylethanolamine and phosphatidylserine, which are asymmetrical distributions in the plasma membrane. A very small proportion of the phosphatidylethanolamine in the outer half had a larger turnover rate than the remaining molecules of the phospholipid in the membrane. Both phosphatidylethanolamine and phosphatidylserine were considered to be present in the membrane as heterogeneous domains, and these domains showed differences in turnover rates. Differences in the fatty acid contents of the phospholipids may be maintained by an asymmetric distribution of enzymes that preferentially incorporate certain fatty acids (see Marinetti and Cattieu, 1982).

Although animal cell membranes derive their phospholipids from the endoplasmic reticulum, a Chinese hamster ovary cell (mutant strain 58) could incorporate, from an exogenous source, at least 50% of the phosphatidylcholine required for membrane biogenesis

(Esko et al., 1982). Esko's study indicates that under certain, perhaps abnormal conditions, animal cells may modify their membrane composition by incorporating lipids from the exogeneous pool. Such an incorporation could be achieved through the fusion of lipid micelles with the plasma membrane (Esko et al., 1982).

15.3. BIOSYNTHESIS AND MATURATION OF MEMBRANE PROTEINS

The biosynthesis of plasma membrane proteins is similar in essential features to that of the secretory proteins (reviews by Lodish et al., 1981; Sabatini et al., 1982). Thus, these two classes of proteins may represent progressive stages of an evolutionary development that first produced the insertion of cytoplasmic proteins into membranes, and later their extrusion and release from the membrane (Sabatini et al., 1982). Studies of simple viral envelope (membrane) proteins have facilitated our current understanding of the biosynthetic pathways of membrane proteins. Vesicular stomatitis virus (VSV) and Sindbis virus contain, respectively, one G protein and two (E1 and E2) glycoproteins in their envelopes. Sindbis virus contains a third protein (core protein) which is associated with the RNA genome. The core proteins and the two membrane glycoproteins are synthesized in a sequence with the core protein first, followed by E1 and a precursor of E2. They are derived by proteolytic cleavage of a nascent chain (see Lodish et al., 1981).

Green et al. (1981) investigated the intracellular transport of two glycoproteins (E1 and p62) of the Semliki Forest virus spike proteins. They used specific antibodies labeled indirectly with ferritin or gold and biochemical fractionation to track the path of the protein after biosynthesis at the rough endoplasmic reticulum. Their data were conclusive that the proteins pass through the Golgi complex before reaching the final destination in the plasma membrane. They acquire oligosaccharide during transport through the Golgi complex. The G protein of the VSV also passes through the Golgi complex (Bergmann et al., 1981).

There may be variations in the time that different membrane proteins reside in the Golgi complex before they appear in the plasma membrane. Using ^{14}c-fucose as a tracer for the plasma membrane glycoproteins in hepatoma tissue culture cells, Doyle et al. (1978) found that only a part of the membrane with fucose-containing glycoproteins is supplied directly to the plasma membrane within three hours after fucosylation, whereas a part of it (~60%) contributes to a large internal pool, the function of which has not been determined. Ach receptors also reside in the Golgi complex for two to three hours before they appear in the plasma membrane (Fambrough and Devreotes, 1978).

While glycosylation of membrane proteins may normally occur concurrently with the insertion of the nascent polypeptide into the endoplasmic reticulum, it does not appear to be a prerequisite for insertion. Using an in vitro system, Rothman et al. (1978) showed that unglycosylated G protein of VSV can be inserted into microsomal membranes.

The first stage in the glycosylation of membrane proteins is the assembly of a 14-sugar oligosaccharide on the lipid carrier, dolichol pyrophosphate (reviewed by Snider, 1982). The oligosaccharide-lipid complex is synthesized by the sequential addition of sugar residues to the lipid in the endoplasmic reticulum. The completed sugar chain is then transferred to the growing polypeptide chain. This peptide-sugar complex is often extensively modified to produce the final glycoprotein. The sugar nucleotides are localized at the cytoplasmic surface of the endoplasmic reticulum but are transferred from the lipid complex to the protein in the lumen of the endoplasmic reticulum. The mechanism by which the sugar residues are transported across the endoplasmic reticulum does not appear to be known (Snider, 1982).

Schmidt and Schlesinger (1979), working on the viral membrane proteins, found that one of the post-translational events involves the addition of fatty acids (1–2 moles per mole of protein) as the membrane glycoproteins move to the cell surface. The significance of this post-translational modification is not known, but it may be required for the transport of the membrane protein to the cell surface. Such a post-translational modification may be a common feature of the membrane protein.

While the secretory proteins and the plasma membrane glycoproteins take apparently similar intracellular pathways, the two classes of proteins may be transported in different vesicles to the cell surface after they leave the Golgi complex. There is a likelihood that a segregation takes place within the Golgi complex and perhaps even earlier within the endoplasmic reticulum, which would be consistent with the existence of functional domains in membranes.

In a multisubunit membrane protein, Sherman et al. (1980) reported that the catalytic (MW 100,000) and the glycoprotein (MW 60,000) subunits of Na^+, K^+-ATPase in the basolateral plasma membrane in kidney epithelial cells are synthesized on the free and bound polysomes, respectively. The catalytic subunit is incorporated into the membranes post-translocationally. Also, the peripheral membrane proteins, for instance, a matrix protein of the VSV, appear to be synthesized as soluble proteins on free polysomes and become associated with the plasma membrane later. Such a transfer of peripheral proteins from the polysomes to the plasma membrane without routing through the endoplasmic reticulum and Golgi complex may be common to this class of proteins. However, a generalization is not warranted at this time as the biogenesis of only a very few peripheral plasma membrane proteins has been studied (see Lodish et al., 1981). The mechanism (or mechanisms) for insertion into and transport to the plasma membrane is not understood. There may be carrier proteins that transport these soluble peripheral proteins to the site of insertion at the plasma membrane.

The maturation of membrane proteins generally requires posttranslational modification, such as glycosylation, linkage with fatty acids, and proteolytic cleavage (Hauri et al., 1979). MacGregor and McElhaney (1981) reported another modification—that nitrate reductase undergoes post-translation modification after insertion into the cytoplasmic membrane in *Escherichia coli*. This enzyme consists of three subunits: A, 142,000; B, 60,000; C, 19,500. The B subunit of the enzyme is synthesized in the form of subunit B′ in the cytoplasm and is converted to the B form found in the functional enzyme. B and B′ forms can be distinguished by a more rapid electrophoretic mobility of the B′ form. According to the signal hypothesis, signal peptide is cleaved off by signal peptidase, but the conversion of B' to B could result from the addition of a peptide, as the electrophoretic mobility becomes slower. However, the precise nature of the modification is not known.

15.4. THE SIGNAL PEPTIDE HYPOTHESIS VERSUS SPONTANEOUS INSERTION INTO THE MEMBRANE

That the secretory proteins, and presumably the membrane proteins, are synthesized with a signal at the N-terminus of a precursor protein was suggested by Milstein et al. (1972). They studied the synthesis of immunoglobulin light chains and found that the precursor protein carried ~15 amino acids extra at the N-terminus and proposed that this short sequence of amino acids might serve as a signaling device for the segregation of secretory proteins that are synthesized on the membrane-bound polysomes.

In contrast to the above signal receptor protein hypothesis, which requires specific protein(s) within the membrane for insertion and transfer across the membrane during protein synthesis, a spontaneous process that does not require specific protein(s) for these events is being advocated (see Bretscher, 1973; Engelman and Steitz, 1981; Wickner, 1980). Engelman and Steitz proposed a helical hairpin hypothesis, according to which a "nascent polypeptide chain folds in an aqueous environment to form an antiparallel pair of helices." This helical hairpin is thought to insert· spontaneously into the hydrophobic lipid bilayer. In cases such as the rbc Band III protein or bacteriorhodopsin in the purple membrane of halobacteria (Chapter 9), where several segments of the protein traverse the lipid bilayer, several hairpin helices may be involved. In this model, one of the helices may be formed by a hydrophobic leader peptide; whether a protein is anchored into the membrane or secreted is determined by the hydrophobic or hydrophilic nature of the two helices in the hairpin. In the case of integral membrane protein, both the helices are hydrophobic, and the hairpin remains inserted in the membrane.

The presence of substantial "β-conformation" has been demonstrated in the case of a few integral membrane proteins studied in detail by the diffraction method, for instance, connexin (on the meridian, maxima are resolved at 4.7Å, characteristic of cross-β'-conformation [Makowski et al., 1982]) and porin (Section 6.4; Schindler and Rosenbusch, 1982). How can an integral membrane protein having predominantly β-structure be inserted into the lipid bilayer? This problem may at first sight appear to be insurmountable for the helical hairpin hypothesis proposed by Engelman and Steitz (1981). However, it can be simply resolved by the consideration of a protein-folding mechanism proposed by Lim (1980). This folding mechanism is based upon the observation that if the entire polypeptide chain of any soluble protein is folded up into an α-helix, the hydrophobic residues would predominantly occupy the helix surface in the form of clusters. Such an α-helix with one hydrophobic cluster is called s-helix, and the polypeptide fragment, in which the s-helix resides, is termed s-fragment. The process of protein folding occurs in three stages. In the beginning, local interactions, in the absence of any long-range interactions, induce the formation of "fluctuating s-helices." The second stage involves the formation of from one to several intermediate "helical globules" from s-helices under the influence of interactions among hydrophobic clusters of s-helices. All or perhaps only a few of the s-helices are either "stabilized or transformed" by the long-range interactions into β-conformations or irregular conformations in the third stage. This protein-folding pathway involves the presence of only s-helices in the beginning, and may eventually lead to the formation of diverse secondary structures, for example, α-helical, β-sheet, and irregular conformations. Therefore, even a β-sheet structure would have to evolve through the primordial s-helices in the folding process, which might facilitate their insertion into the lipid bilayer, as explicated in the helical hairpin hypothesis. Further studies should elucidate the molecular mechanism (or mechanisms) of protein insertion into and across the membranes, the role of signal peptides, and possible spontaneous insertion.

Braell and Lodish (1982) studied the biosynthesis of Band III glycoprotein of the rbc plasma membrane. They used the mRNA from the erythroid precursor cells from the spleen of anemic mice in an in vitro cell-free system containing wheat germ and reticulocyte lysate. Microsomal membranes from dog pancreas were added to the in vitro system. Their results indicate that Band III protein is cotranslationally inserted into microsomes. Even when the microsomes were added after nearly one half of the NH_2-terminal (40,000–45,000 MW protein) had been synthesized, the transmembrane insertion into microsomal membrane is near the middle of the protein; there was no cleavage of a signal peptide. This contrasts with the viral integral membrane proteins and many other integral membrane proteins and secretory proteins that have an NH_2-terminal secretion that is generally cleaved off during or just after translation (De Villers-Thiery et al., 1975; Lodish et al., 1981).

Anderson et al. (1982) demonstrated that one of the subunits of the Ach receptor, that is, the δ subunit, is synthesized with an NH_2-terminal signal sequence that is cleaved on integration into the membranes of the endoplasmic reticulum. This transient signal sequence has 21 residues. The integration of integral membrane proteins (and secretory proteins) is thought to require a signal recognition protein (Walter and Blobel, 1981). This signal recognition protein (11S) appears to function in polysome recognition and binding to the microsomal membrane, events that are a prerequisite for chain translocation. In a cell-free protein-synthesizing system in the presence of signal recognition protein, but in the absence of microsomal membranes, the synthesis of the transmembrane glycoprotein subunit of the Ach receptor was found to be inhibited. The inhibition was abolished with the addition of the microsomal membranes. The specific role of the signal recognition protein in translocation and integration of the integral membrane proteins (and secretory proteins) remains to be understood.

15.5. SITES OF PROTEIN INSERTION INTO THE PLASMA MEMBRANE

Membrane proteins may be initially inserted in the plasma membrane at sites other than where they manifest their functions. Such an insertion requires mobility from the initial incorporation to the final destination. As an example, the integral membrane glycoproteins of the Semliki Forest virus (SFV) that make up the spikes on the envelope (E1, E2, and E3 proteins) are inserted into the plasma membrane of the host cell at sites different than where the viruses bud off (Simons et al., 1982). Another instance of this type, the growth cone in neurite outgrowth, provides a rapid expansion of the neuron plasma membrane (Pfenninger, 1981). The plasma membrane of the growth cone differs from that of the neuron cell body in lectin binding sites and is relatively devoid of IMPs. Pfenninger employed a variety of approaches for such an investigation, for instance, pulse-chase experiments with lectins, measurement of the density of IMPs in the plasma membrane of the growing neurites as a function of time and distance from the cell body, and intracellular transport of tritiated phospholipid precursor (^{3}H-glycerol). His data led him to conclude that the growth of the neurite takes place by fusion of IMP-free vesicles with the plasma membrane. These vesicles carry newly synthesized glycoconjugates and are transported from the cell body. (These glycoconjugates are either glycolipids or small glycopeptides.) Furthermore, the IMPs are inserted into the plasma membrane of the cell body and reach the growth cone by lateral diffusion. These interpretations of membrane biosynthesis require further close scrutiny. Feldman et al. (1981) came to the conclusion that the new membrane is added in the region of the growing tip in the neurite in cultured goldfish retina.

In contrast to the viral spike proteins and perhaps neurite growth cone plasma membrane, the α-BGT binding sites (Ach receptors) in the plasma membrane of skeletal muscle and perhaps many other integral membrane proteins are inserted in the plasma membrane as functional entities (Carbonetto and Fambrough, 1979; Pumplin and Fambrough, 1982).

16

Some Biomedical Applications of the Plasma Membrane

A search for cell surface molecules involved in cell-to-cell and cell-to-environment interactions, and in cell specificity is particularly relevant to nerve–target interaction, to histogenesis, and to the understanding of the role of specific membrane components. In this respect, sympathetic neurons in cultures provide a useful model system since these neurons are capable of forming adrenergic or cholinergic synapses, which can be controlled through the use of various factors. Depolarization with elevated K produces adrenergic synapses, whereas conditioned heart cell medium produces cholinergic synapses (Braun et al., 1981). Three glycoproteins (labeled C55, C for cholinergic; A82, A for adrenergic; and A155, the numbers indicate the apparent MW) were identified that are apparently related to the induction of synthesis and accumulation of acetylcholine or catecholamines. The relative amounts of C55 were 4.95% in cultures grown in heart conditioned medium and 0.27% from cultures grown in elevated K medium. A82 was 2.34% in adrenergic cultures and 0.56% in cholinergic cultures. C55 and A155 are accessible to labels from outside the cell (namely neuraminidase, lactoperoxidase, and galactose oxidase) and thus thought to be exposed at the cell surface. A82 is considered to be a secretory glycoprotein and is thereby released into the culture medium. Further studies of this nature should advance our understanding of the nature of specific interactions between cells, particularly the intercellular communication between various types of neurons.

Another cell surface specific molecule that appears to be involved in cell-to-cell adhesion in the nervous system is the cell adhesion molecule (MW 140,000; Buskirk et al., 1980; Thiery et al., 1977). Antibodies raised to this molecule have been reported to disrupt histogenesis of the developing chick retina in organ culture.

It is of further interest that the malignant cells have been reported to contain a glycoprotein (~MW 90,000) in their plasma membrane that is "absent or greatly reduced" on the surface of nonmalignant cells. Using [14$_C$-] glucosamine, Atkinson and Bramwell (1981) reported that there is no difference in the malignant and nonmalignant cells in the enzymes concerned with the conversion of the precursor into sugars. They therefore speculated that the alterations in the malignant cells may depend on the insertion, the configuration, or the amount of glycoprotein on the cell surface.

Jones and Witkowski (1981) reported that the skin fibroblasts from Duchenne muscular dystrophy patients manifested a reduced intercellular adhesiveness from normal cell suspensions, which suggests an alteration in the surface properties of dystrophic cells. Also, in the dystrophic mouse the Schwann cells fail to develop normally in the nerve roots and this abnormality has been hypothesized to be related to a deficiency in the extracellular material surrounding the Schwann cell (Bunge and Bunge, 1978; Fig. 16.1).

Muscular dystrophies are well known to be heterogeneous disorders characterized by muscle weakness and degeneration. The sex-linked Duchenne muscular dystrophy has been extensively studied, but the mechanism of the disorder is not known (Witkowski and Jones, 1981). Various animal model systems are being investigated,

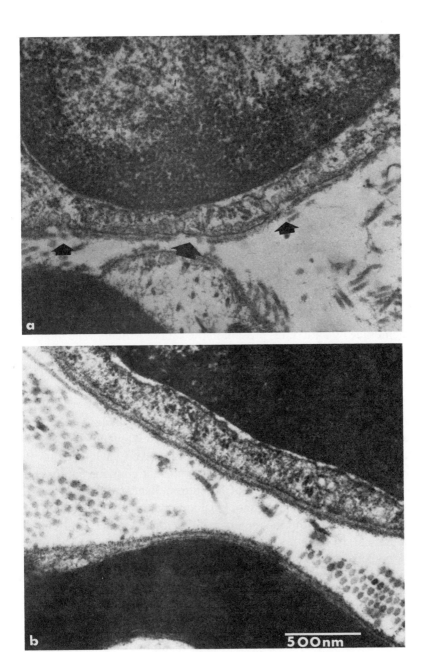

Figure 16.1 Electron micrographs of sciatic nerve in section showing basement membrane from a dystrophic (a) and normal (b) mouse. Arrows indicate a discontinuous basement membrane in dystrophic animal which may be a result of malfunctioning of the Schwann cells.

163

and current efforts have been extended to include nonmuscle cells: In a number of cells, for instance, fibroblasts, lymphocytes, and red blood cells, several biochemical characteristics of the plasma membrane have been found to differ from their normal counterparts (Pimplikar and Malhotra, 1983; Rosenmann et al., 1982; Wilkerson et al., 1978; Witkowski and Jones, 1981). The emphasis therefore is on the study of membrane abnormality.

Another well-known disorder related to the cell surface molecules is *Myasthenia gravis,* which has been extensively reviewed by Lewin (1981; see Grob, 1981). *Myasthenia gravis* is characterized by the weakening of muscles. Earlier studies suggested that this disease resulted from a defect in the release of the transmitter acetylcholine from the nerve endings. However, more recent studies by Lindstrom, Lennon, Drachman, and others indicate that the disorder results from disruption of the Ach receptors on the postsynaptic membrane by antibodies present in the sera of patients. The molecular events that cause the disorder are gradually being understood. It appears that most antibodies to the Ach receptor bind at sites on the receptors other than the Ach binding site. Antibodies also cross-link two Ach receptors (antigenic modulation). They also interact with a component of the immune system called *the complement.* Whatever the mechanism of interaction between the Ach receptor antibodies and the postsynaptic membrane, there is a severe deficiency of the functional receptors in the membrane, since it contains barely one-third of the normal density of receptors. Recent advances in the techniques of producing monoclonal antibodies should facilitate the understanding of the machinery of *Myasthenia gravis* and consequently assist attempts to control the disease. The available data on monoclonal antibodies suggest that many of the antibodies bind only to the receptors against which they are produced. However, about half of the isolated number of 70 monoclonal antibodies cross-react with receptors of other species. The binding site of these antibodies has been reported to be on the α-subunit of the Ach receptor, but it is not the site where Ach binds. Similar antibody binding sites are detectable in the receptors of *Torpedo,* eel, calf muscle, and human muscle, which are highly conserved through evolution. It might prove to be an important functional region and a prime target in the autoimmune response of *Myasthenia gravis.*

Immunosuppressive drugs are administered over prolonged periods as treatment for *Myasthenia gravis* patients. However, it was discovered that dimethyl sulphoxide (DMSO), which was initially used as a solvent for one of the drugs, produced a rapid reduction in the anti-AchR antibody in an experimental case of autoimmune *Myasthenia gravis* induced in rat by immunization with Ach receptor, and this action of DMSO was sustained even after its use was discontinued. The mode of action of DMSO remains to be elucidated, though it has been conjectured that it might be effective indirectly in the production of antibodies by the lymphoid cells (Pestronk and Drachman, 1980a). (A detailed account of the Ach receptor is given in Chapter 7.)

In the genetic disease known as familial hypercholesterolemia, abnormal cholesterol homeostasis has been associated with a series of defects in the receptor for LDL particles that provide mammalian cells with cholesterol. Using cultured fibroblasts from normal humans and from patients, Brown and Goldstein (1979a, b) found that in one type the fibroblasts were lacking LDL receptors on the plasma membrane, which prevented uptake of LDL particles. This resulted in alterations in the intracellular regulation of cholesterol biosynthesis. In another type, the LDL receptors were greatly reduced in number. In the third type, the number of LDL receptors was normal, but the LDL could not be internalized. This defect was traced to the inability of the LDL receptors to laterally move into the coated pits where they are normally endocytosed (Chapter 14). It is recalled that the LDL particles (20–25 nm) in human plasma contain cholesterol in esterified form in their core, surrounded by a coat of phospholipid, free cholesterol, and a protein. After endocytosis, the LDL particles fuse with the lysosomes where the coat protein and cholesteryl esters are hydrolysed. The freed cholesterol is made available for the biosynthesis of membranes. Related to the cholesterol metabolism are other cellular processes such as LDL receptor biosynthesis, reesterification of excess of cholesterol, and biosynthesis of cholesterol, which are controlled by the intracellular level of cholesterol (Brown and Goldstein, 1979a, b). In confirmation of the receptor-mediated endocytosis of LDL particles in fibroblasts, Anderson et al. (1981) found that the human epithelioid carcinoma cells (A-431) had an increased number of LDL receptors compared to that in the fibroblasts. These receptors were distributed over the entire plasma membrane. The LDL endocytosis was, however, inefficient, a condition traced, again, to the inability of the receptors to be incorporated into the coated pits.

A number of abnormalities pertaining to the myelin sheath in strains of mutant mice (Bunge, 1980) provide important advances in our understanding of the organization of myelin. For example, in a strain of mouse described as Twicher, there is a progressive accumulation of cerebroside as the amount of myelin declines. This abnormality results from a deficiency of the enzyme that breaks down cerebroside to ceramide and galactose. This pathological state is reminiscent of the human degenerative disorder known as Krebb's disease, which is characterized by the accumulation of macrophages with crystalloid inclusions in the myelin-deficient areas. In the Jimpy mouse, there appears to be little formation of myelin in the central nervous system, as a myelin precursor has been reported to accumulate in the cell without being incorporated into the myelin. This abnormality is reminiscent of the Pelizaeus-Merzbacher disease in man, which is often fatal during the first three or four years of life.

In another mouse, the Trembler, there is an abnormally sparse myelination (see Matthieu et al., 1981) and a deposition of connective tissue constituents that eventually enlarge the peripheral nerve trunks. Such nerves manifest a slower conduction velocity and a reduced diameter of the axon. This type of disorder is manifested in patients with Charcot-Marie-Tooth disease, which is a hereditary disease with peripheral nerve hypertrophy. The Shiverer mutant specifically lacks the extrinsic basic protein of the myelin in the central nervous system—less than 3% of the normal content is detected (Dupouey et al., 1979). The structure of the myelin sheath in the central nervous system was found to be altered (a compact sheath is not formed [Privat et al., 1979]), whereas the myelin in the peripheral nervous system, which also lacks the basic protein, appeared normal by electron microscopy and X-ray diffraction (Kirschner and Ganser, 1980). There may be apparent differences in the role that the basic protein serves in myelin in the central nervous system and the peripheral nervous system (Bunge, 1980). Even in normal animals the peripheral nervous system manifests a large variation in the basic protein in myelin, from 6% in rat sciatic nerve to 22% in rabbit sciatic nerve, which is lower than the 25–50% in the central nervous system. Such a variation in the amount of the basic protein might make one question a uniform structural role in the formation of the myelin sheath. Basic proteins have an apparent MW of ~18,000–20,000 (see Kirschner and Ganser, 1980).

It should be of interest that antibody to basic myelin protein has been detected in extracts from the brain of multiple sclerosis patients. Multiple sclerosis is a well-known disease of the central nervous system that produces loss of the myelin sheath (Bernard et al., 1981).

The merozoites of the human malarial parasite *Plasmodium falciparum* enter the red blood cell, where they grow and multiply. It appears that glycophorin, which is a major sialoglycoprotein of the plasma membrane of the erythrocyte, serves as a receptor for the parasite (Perkins, 1981; reviewed by Wilson, 1982). Therefore, entry of the parasite into the red blood cell could be inhibited or prevented by experimental blocking of the receptor molecules by antibodies. The attachment of the parasite to the red blood cell takes place at the apical end only of the merozite (see Wilson, 1982). This is a prerequisite for its entry into the red blood cell. One should therefore be able to raise monoclonal antibodies to block invasion of the red blood cells by the malarial parasite. Further, *Plasmodium vivax* has been reported to be absent from West Africa where most of the population is negative in respect of human Duffy blood group antigens. Human red blood cells from this population are resistant to invasion by the Simian parasite *Plasmodium knowleski*. Therefore, there may be several antigenic molecules on the surface of the red blood cell for recognition and attachment of the parasite, and molecular dissection of these sites should enhance our understanding of the interactions between the two cells. The attachment of the parasite to the plasma membrane of the erythrocyte produces alterations in the structure of the membrane that can be followed by freeze-fracture and, in thin sections, by electron microscopy (Aikawa et al., 1981).

17

Concluding Remarks

In view of the vital and highly dynamic interactions between the plasma membrane and its intracellular and extracellular environments, the current emphasis on studies on natural membranes (in situ and in vitro) is likely to persist, particularly with physiological and developmental stress. Such studies are attractive in neurobiology, and the nervous system can provide a valuable model for studies of the various functional domains that exist in an otherwise physically continuous membrane. The nervous system also undergoes developmental modification of its membranes, which are discussed in a recent volume of the Neurosciences Research Program Bulletin (Rakic and Goldman-Rakic, 1982). The structure of the myelin sheath undergoes changes during maturation (Franks et al., 1982), and the course of appearance of neuronal cell surface and synaptic vesicle antigens has been investigated by using monoclonal antibodies (Greif and Reichardt, 1982).

Studies of nervous, as well as nonnervous cells, have been useful for the investigation of intracellular factors involved in membrane fusion. Besides calcium, GTP has been implicated in the fusion of intracellular membrane vesicles during glycosylation (Paiement et al., 1980).

Both nervous and various nonnervous cells are also well suited for studies of the role of the cytoskeleton in regulating the activities of the plasma membrane in normal and pathological tissues. Researchers have already reported results on the influence of extracellular structures, such as the basement membrane, on the functioning of the plasma membrane. For instance, a discontinuity of the basement membrane surrounding Schwann cells in nerves of dystrophic mouse (Fig. 16.1) probably results from a defect in Schwann cell function (Madrid et al., 1975). Whether this abnormality of the basement membrane is a general feature of dystrophic animals is not known, but it is now recognized that dystrophy is a group of heterogeneous abnormalities in which various tissues besides muscle are affected (Rosenmann et al., 1982). Another disease of the nervous system is multiple sclerosis which is characterized by demyelination, which affects nerve conduction (Arnason, 1982). It would be of interest to know the nature of the abnormality in the functioning of the oligodendrocytes, which normally produce the myelin sheath in the CNS.

The role of antibodies in the degradation of Ach receptors in the autoimmune disease *Myasthenia gravis* is now recognized (Chapter 16; Grob, 1981). The inability of the LDL receptors to laterally move into the coated pits in the plasma membrane results in an abnormality in cholesterol metabolism (Chapter 14; Brown and Goldstein, 1979a, b). An understanding of the biosynthetic process of membranes should facilitate advances in developing strategies to handle some of these and other membrane-related abnormalities.

Mechanisms by which parasites, venoms, drugs, and antibodies interact with the plasma membrane are being extensively investigated. The entry of the malarial parasite into rbc and the action of cholera toxin were mentioned in Chapter 6 and Section 11.3, respectively. The binding of antibodies to antigenic determinants on the cell surface often results in cytolysis. The mechanism by which the complement system, consisting of a group of proteins, acts sequelly to mediate the binding of antibody and antigen is being explored at the level of single-channel currents (Jackson et al., 1981; Ramm et al., 1982). Of the more than 20 proteins in the complement system, five (C5, C6, C7, C8, and C9) are involved in binding to the membrane and in cytolysis. These proteins self-assemble into a macromolecular membrane-attack complex upon enzymatic cleavage when a single peptide bond is initiated. C9 (MW 71,000), which is a water-soluble protein, has been found to undergo polymerization and forms a transmembrane protein channel (diameter ∼11 nm). Tschopp et al. (1982) found, by CD spectroscopic analysis, an increase in the β-sheet conformation of protein (38%) and a slight decrease in the α-helix contents (22%) of the polymerized C9. This, compared to the 32% of β-sheet and the 24% α-helix in monomeric C9, indicates a conformation change upon polymerization. Besides advancing the understanding of the mechanism of cell lysis, such studies reveal how water-soluble proteins might undergo conformational changes and facilitate insertion into lipid bilayers (Tschopp et al., 1982).

The bee venom contains melittin, which consists of 26 amino acid residues at the C terminal. Six of these are positively charged and ten are polar. It is water-soluble, forms tetramers, and lyses cells by integrating into membranes. Crystallographic studies indicate that the polypeptide has two α-helical segments and has the shape of a bent α-helical rod in which the hydrophobic side-chains are on the inner surface, whereas the hydrophilic side-chains are on the outer surface. Melittin appears to be parallel to the membrane on its surface, and as the surface area of the membrane increases on one side, it leads to the formation of lipid pores stabilized by melittin. These lipid pores eventually increase in size, and the membrane disintegrates into flat sheets (Terwilliger et al., 1982).

Molecular cloning of cDNA coding for one of the subunits, that is, γ-subunit (MW 60,000), of the Ach receptor from *Torpedo*'s electric organ has been achieved. The cDNA was found to hybridize to a 2100 base mRNA in the electric organ of *Torpedo* but not in brain (Ballivet et al., 1982), which indicates differences in the Ach receptors in the two locations.

A highly purified TTX binding protein (sodium channel) has been isolated from the electric eel *Electrophorus*. It has MW 260,000 and forms clusters of rod-shaped particles ($\sim 4 \times 17$ nm) when examined by electron microscopy after negative staining (Ellisman et al., 1982). It is apparent that the isolation and characterization of the sodium channel protein is nearing completion. Monoclonal antibodies are available that could establish homologies between TTX-sensitive and TTX-insensitive sodium channels (Lawrence and Catterall, 1981).

Studies on intercellular coupling are important for understanding the mechanisms through which cells of a tissue coordinate their normal cellular activities during development and in the adult life of an animal. Beyond the participation of calcium and/or pH and cAMP (Section 6.3), not much seems to be known about the information exchanged between cells nor the mechanisms of information exchange. How different are the gap junctions in various vertebrates and invertebrates could be revealed, in part, by the application of monoclonal antibodies.

Three-dimensional image reconstructions of membrane proteins have been facilitated by low-dose, high-resolution electron microscopy. Only a few membrane proteins have been thus far subjected to such analyses because of the problem of getting the ordered (crystalline) specimen required for these studies (Henderson, 1980). Yet such studies, when they are feasible, could reveal important structural similarities between various membrane proteins, particularly when combined with protein conformational data that can be obtained from CD spectroscopy (Jap et al., 1982).

Underlying the rather simple-looking structure of membranes is the vast diversity of chemical constituents in natural membranes indicative of the diverse roles these membranes serve in the lives of cells. These constituents are synthesized, transported, and assembled into functional domains that interact with other membranous domains as well as with their often-changing environments in normal and pathological cells. How cells of an organism achieve these functions is as yet incompletely understood.

References

Agard, D. A. and Stroud, R. M. *Biophys. J.* **37:**589–602 (1982).

Ahkong, Q. F.; Fisher, D.; Tampion, W.; and Lucy, J. A. *Nature,* London **253:**194–195 (1975).

Aikawa, M.; Miller, L. H.; Rabbege, J. R.; and Epstein, N. *J. Cell Biol.* **91:**55–62 (1981).

Anderson, D. J.; Walter, P.; and Blobel, G. *J. Cell Biol.* **93:**501–506 (1982).

Anderson, E. and Albertini, D. F. *J. Cell Biol.* **71:**680–686 (1976).

Anderson, J. M. *J. Biol. Chem.* **254:**939–944 (1979).

Anderson, R. G. W.; Brown, M. S.; and Goldstein, J. L. *J. Cell Biol.* **88:**441–452 (1981).

Anderson, R. G. W.; Brown, M. S.; Beisiegel, U.; and Goldstein, J. L. *J. Cell Biol.* **93:**523–531 (1982).

Anglister, L. and Silman, I. *J. Mol. Biol.* **125:**293–311 (1978).

Anholt, R.; Linstrom, J.; and Montal, M. *J. Bio. Chem.* **256:**4377–4387 (1981).

Aranson, B. G. W. *Hospital Practice* **17:**81–89 (1982).

Atkinson, M. A. L. and Bramwell, M. E. *J. Cell Sci.* **48:**147–170 (1981).

Axelrod, D.; Ravdin, P.; Koppel, D. E.; Schlessinger, J.; Webb, W. W.;

Elson, E. L.; and Podleski, T. R. *Proc. Natl. Acad. Sci.,* U.S. **73**:4594–4598 (1976).

Axelrod, J. *Neurosci. Res. Prog. Bull.* **20**:327–338 (1982).

Axelsson, J. and Thesleff, S. *J. Physiol.* **147**:178–193 (1959).

Bacou, F.; Vigneron, P.; and Massoulie, J. *Nature,* London **296**:661–664 (1982).

Bader, D. *J. Biochem.* **255**:8340–8350 (1980).

Baenziger, J. U. and Fiete, D. *J. Biol. Chem.* **257**:6007–6009 (1982).

Baker, P. F. and Knight, D. E. *Phil. Trans. R. Soc.,* London **B 296**:83–103 (1981).

Bakke, A. C. and Lerner, R. A. *Subcell. Biochem.* **8**:75–122 (1981).

Ball, E. H. and Singer, S. J. *Proc. Natl. Acad. Sci.,* U.S. **78**:6986–6990 (1981).

Ballivet, M.; Patrick, J.; Lee, J.; and Heinemann, S. *Proc. Natl. Acad. Sci.,* U.S. **79**:4466–4470 (1982).

Banks, P. *Nature,* London **287**:12–13 (1980).

Bareis, D. L.; Hirata, F.; Schiffmann, E.; and Axelrod, J. *J. Cell Biol.* **93**:690–697 (1982).

Barnard, E. A. and Dolly, J. O. *Trends in Neurosci.* **5**:325–327 (1982).

Barrantes, F. J. *J. Cell Biol.* **92**:60–68 (1982).

Barritt, G. J. *Trends Biochem. Sci.* **6**:322–325 (1981).

Bearer, E. L. and Friend, D. S. *J. Cell Biol.* **91**:266a (1981).

Bennett, M. V. L. and Goodenough, D. A. *Neurosci. Res. Prog. Bull.* **16**:373–486 (1978).

Bennett, M. V. L.; Aljure, E.; Nakajima, Y.; and Pappas, G. D. *Science* **141**:262–264 (1963).

Bennett, M. V. L.; Spray, D. C.; and Harris, A. L. *Trends Neurosci.* **4**:159–163 (1981).

Bennett, V. and Stenbuck, P. J. *Nature,* London **280**:468–473 (1979).

Bentzel, C. J.; Hainau, B.; Edelman, A.; Anagnostopoulos, T.; and Benedetti, E. L. *Nature,* London **264**:666–668 (1976).

Bentzel, C. J.; Hainau, B.; Ho, S.; Hui, S. W.; Edelman, A.; Anagnosto-poulos, T.; and Benedetti, E. L. *Am. J. Physiol.* **239**(Cell Physiol. 8):c75–c89 (1980).

Berdan, R. C. and Caveney, S. *J. Cell Biol.* **91**:99a (1981).

Bergman, K.; Burke, P.; Cerda-Olmedo, E.; David, C.; Delbrück, M.;

Foster, K.; Goodell, E.; Heisenberg, M.; Meissner, G.; Zalokar, M.; Dennison, D.; and Shropshire, W., Jr. *Bact. Rev.* **33**:99–157 (1969).

Bergmann, J. E.; Tokuyaku, K. T.; and Singer, S. J. *Proc. Natl. Acad. Sci., U.S.* **78**:1746–1750 (1981).

Bernard, C. C. A.; Randell, V. B.; Horvath, L. B.; Carnegie, P. R.; and Mackay, I. R. *Immunology* **43**:447–458 (1981).

Bernhardt, R. and Matus, A. *J. Cell. Biol.* **92**:589–593 (1982).

Berwald-Netter, Y.; Martin-Moutot, N.; Koulakoff, A.; and Courand, F. *Proc. Natl. Acad. Sci.*, U.S. **78**:1245–1249 (1981).

Betz, H. and Changeux, J-P. *Nature,* London **278**:749–752 (1979).

Blau, L. and Bittman, R. *J. Biol. Chem.* **253**:8366–8368 (1978).

Blaurock, A. E. and King, G. I. *Science* **196**:1101–1104 (1977).

Blaurock, A. E. and Stoeckenius, W. *Nature,* London **233**:152–155 (1971).

Bloch, R. J. *J. Cell Biol.* **82**:626–643 (1979).

Blundell, T. L. and Johnson, L. N. *Protein Crystallography.* Academic Press, New York, 1976.

Boggs, J. M. and Moscarello, M. A. *Biochim. Biophys. Acta* **515**:1–21 (1978).

Bonner, J. T. *J. Exp. Zool.* **106**:1–26 (1947).

Borochov, H. and Shinitzky, M. *Proc. Natl. Acad. Sci.*, U.S. **73**:4526–4530 (1976).

Boulan, E. R. and Sabatini, D. D. *Proc. Natl. Acad. Sci.*, U.S. **75**:5071–5075 (1978).

Boyles, J. and Bainton, D. F. *Cell* **24**:905–914 (1981).

Braell, W. A. and Lodish, H. F. *Cell* **28**:23–31 (1982).

Brandt, B. L.; Hagiwara, S.; Kidokoro, Y.; and Miyazaki, S. *J. Physiol.* **263**:417–439 (1976).

Branton, D. *Proc. Natl. Acad. Sci.*, U.S. **55**:1048–1056 (1966).

Branton, D.; Bullivant, S.; Gilula, N.; Karnovsky, M.; Moor, H.; Muhlethaler, K.; Northcote, D.; Packer, L.; Satir, B.; Satir, P.; Speth, V.; Staehlin, L.; Steere, R.; and Weinstein, R. *Science* **190**:54–56 (1975).

Braun, S. J.; Sweadner, K. J.; and Patterson, P. H. J. *J. Neurosci.* **1**:1397–1406 (1981).

Breimer, M. E.; Hansson, G. C.; Karlsson, K-A.; and Leffler, H. *J. Biol. Chem.* **257**:557–568 (1982).

Bretscher, A. and Weber, K. *J. Cell Biol.* **86**:335–340 (1980).

Bretscher, M. S. *Science* **181**:622–629 (1973).

Bretscher, M. S. *Nature,* London **231**:229–232 (1971).

Bretscher, M. S. and Raff, M. S. *Nature,* London **258**:43–49 (1975).

Bretscher, M. S.; Thompson, J. N.; and Pearse, B. M. F. *Proc. Natl. Acad. Sci.,* U.S. **77**:4156–4159 (1980).

Bridgman, P. C. and Nakajima, Y. *Proc. Natl. Acad. Sci.,* U.S. **78**:1278–1282 (1981).

Brismar, T. *Trends Neurosci.* **5**:179–181 (1982).

Brisson, A. *Ninth Intern. Cong. E. M.* **2**:180–181 (1978).

Brockes, J. P. and Hall, Z. W. *Biochemistry* **14**:2100–2106 (1975).

Brown, M. S. and Goldstein, J. L. *Proc. Natl. Acad. Sci.,* U.S. **76**:3330–3337 (1979a).

Brown, M. S. and Goldstein, J. L. *Harvey Lectures,* Ser. **73**:163–201 (1979b).

Brown, M. S.; Goldstein, J. L.; Krieger, M.; Ho, Y. K.; and Anderson, R. G. W. *J. Cell Biol.* **82**:597–613 (1979).

Browne, C. L.; Wiley, H. S.; and Dumont, J. N. *Science* **203**:182–183 (1979).

Brunner, J.; Hauser, H.; Braun, H.; Wilson, K. J.; Wacker, H.; O'Neill, B.; and Semenza, G. *J. Biol. Chem.* **254**:1821–1828 (1979).

Bullivant, S. *Phil. Trans. Roy. Soc. Lond.* **B 268**:5–14 (1974).

Bunge, R. P. *Nature,* London **286**:106–107 (1980).

Bunge, R. P. and Bunge, M. B. *J. Cell Biol.* **78**:943–950 (1978).

Burden, S. J.; Sargent, P. B.; and McMahan, U. J. *J. Cell Biol.* **82**:412–425 (1979).

Burghardt, R. C. *J. Cell Biol.* **91**:102a (1981).

Buskirk, D. R.; Thiery, J. P.; Rutishauser, U.; and Edelman, G. M. *Nature,* London **285**:488–489 (1980).

Capaldi, R. A. *Membrane Proteins and their Interactions with Lipids.* Ed. R. A. Capaldi, Marcel Dekker, New York **1**:1–19, 1977.

Carbonetto, S. and Fambrough, D. M. *J. Cell Biol.* **81**:555–569 (1979).

Cartaud, J.; Benedetti, E. L.; Sobel, A.; and Changeux, J-P. *J. Cell Sci.* **29**:313–337 (1978).

Cartaud, J.; Sobel, A.; Rousselet, A.; Devaux, P. F.; and Changeux, J-P. *J. Cell Biol.* **90**:418–426 (1981).

Caspar, D. L. D.; Goodenough, D. A.; Makowski, L.; and Phillips, W. C. *J. Cell Biol.* **74:**605–628 (1977).

Catterall, W. A. *J. Neurosci.* **1:**777–783 (1981).

Chang, C. C. and Lee, C. Y. *Arch. Int. Pharmacodyn. Ther.* **144:**241–257 (1963).

Changeux, J-P. and Dennis, S. G. *Neurosci. Res. Prog. Bull.* **20:**269–426 (1982).

Chapman, D. *Membrane Structure and Function.* Ed. E. E. Bittar, Wiley, New York **1:**103–152 (1980).

Chapman, D.; Gomez-Fernandez, J. C.; and Goni, F. M. *Trends Biochem. Sci.* **7:**67–70 (1982).

Chapman, D.; Williams, R. M.; and Ladbrooke, B. D. *Chem. Phys. Lipids* **1:**445–475 (1967).

Cheng, H. and Farquhar, M. G. *J. Cell Biol.* **70:**660–670 (1976a).

Cheng, H. and Farquhar, M. G. *J. Cell Biol.* **70:**671–684 (1976b).

Cheung, W. Y. *Sci. Amer.* **246:**62–70 (1982a).

Cheung, W. Y. *Fed. Proc.* **41:**2253–2257 (1982b).

Chiu, S. Y. and Ritchie, J. M. *Nature,* London **284:**170–171 (1980).

Chiu, S. Y. and Ritchie, J. M. *J. Physiol.,* London **313:**415–437 (1981).

Chou, P. Y. and Fasman, G. D. *Ann. Rev. Biochem.* **47:**251–276 (1978).

Christian, C. N.; Daniels, M. P.; Sugiyama, H.; Vogel, Z.; Jacques, L.; and Nelson, P. G. *Proc. Natl. Acad. Sci.,* U.S. **75:**4011–4015 (1978).

Chubb, I. W. *Trends Neurosci.* **3:**12–13 (1980).

Citri, Y. and Schramm, M. *Nature,* London **284:**297–300 (1980).

Cohen, C. M. and Branton, D. *Trends Biochem. Sci.* **6:**266–268 (1981).

Cohen, C. M. and Foley, S. F. *J. Cell Biol.* **91:**305a (1981).

Colquhoun, D. and Sakmann, B. *Nature,* London **294:**464–466 (1981).

Cone, R. A. *Nature New Biol.* **236:**39–43 (1972).

Conti-Tronconi, B. M. and Raftery, M. A. *Ann. Rev. Biochem.* **51:**491–530 (1982).

Conti-Tronconi, B.; Tzartos, S.; and Lindstrom, J. *Biochem.* **20:**2181–2191 (1981).

Costello, M. J.; Ting-Beall, H. P.; and Robertson, D. *Biophys. J.* **37:**276a (1982).

Craig, S. W. and Pollard, T. D. *Trends Biochem. Sci.* **7:**88–92 (1982).

Cuatrecasas, P. *Biochemistry* **12**:3558–3566 (1973).

Cullis, P. R. and Hope, M. J. *Nature,* London **271**:672–674 (1978).

Daniel, E. E.; Daniel, V. P.; Duchon, G.; Garfield, R. E.; Nichols, M.; Malhotra, S. K.; and Oki, M. *J. Memb. Biol.* **28**:207–239 (1976).

Danielli, J. F. and Davson, H. *J. Cell. Comp. Physiol.* **5**:495–508 (1935).

Das, M. and Fox, F. *Proc. Natl. Acad. Sci.,* U.S. **75**:2644–2648 (1978).

Davis, R. L. and Kiger, J. A., Jr. *J. Cell Biol.* **90**:101–107 (1981).

De Chastellier, C. and Ryter, A. *Biol. Cell* **40**:109–118 (1981).

de Latt, S. W.; Van Der Saag, P. T.; Elson, E. L.; and Schlessinger, J. *Proc. Natl. Acad. Sci.,* U.S. **77**:1526–1528 (1980).

De Mello, W. C. *Membrane Structure and Function.* Ed. E. E. Bittar, Wiley, New York **3**:127–170 (1980).

De Villers-Thiery, A.; Kindt, T.; Scheele, G.; and Blobel, G. *Proc. Natl. Acad. Sci.,* U.S. **72**:5016–5020 (1975).

Dehlinger, P. J.; Jost, P. C.; and Griffith, O. H. *Proc. Natl. Acad. Sci.,* U.S. **71**:2280–2284 (1974).

Dennis, S. G. *Neurosci. Res. Prog. Bull.* **20**:350–353 (1982).

Devreotes, P. N. and Fambrough, D. M. *Cold Spring Harbor Symp. Quant. Biol.* **40**:237–251 (1976).

Doyle, D.; Baumann, H.; England, B.; Friedman, E.; Hou, E.; and Tweto, J. *J. Biol. Chem.* **253**:965–973 (1978).

Drachman, D. B.; Pestronk, A.; and Stanley, E. F. *Adv. Cytopharm.* **3**:237–244 (1979).

Drachman, D. B.; Stanely, E. F.; Pestronk, A.; Griffin, J. W.; and Price, D. L. *J. Neurosci.* **2**:232–243 (1982).

Dragsten, P. R.; Blumenthal, R.; and Handler, J. S. *Nature,* London **294**:718–722 (1981).

Dudai, Y.; Herzerg, M.; and Silman, I. *Proc. Natl. Acad. Sci.,* U.S. **70**:2473–2476 (1973).

Dunlap, K. *J. Physiol.* **27**:119–133 (1977).

Dupouey, P.; Jacque, C.; Bourre, J. M.; Cesselin, A.; Privat, A.; and Baumann, N. *Neurosci. Lett.* **12**:113–118 (1979).

Edelman, G. M. *Science* **192**:218–226 (1976).

Edmonds, D. T. *Trends Biochem. Sci.* **6**:92–94 (1981).

Ellisman, M. H.; Agnew, W. S.; Miller, J. A.; and Levinson, S. R. *Proc. Natl. Acad. Sci.,* U.S. **79**:4461–4465 (1982).

Engelman, D. M. and Steitz, T. A. *Cell* **23**:411–422 (1981).

Engelman, D. M.; Goldman, A.; and Steitz, T. A. *Meth. Enzymol.* **88**:81–88 (1982).

Engelman, D.; Henderson, R.; McLachlan, A.; and Wallace, B. *Proc. Natl. Acad. Sci.*, U.S. **77**:2023–2027 (1980).

Engleman, D. M. and Zaccai, G. *Proc. Natl. Acad. Sci.*, U.S. **77**:5894–5898 (1980).

Erickson, H. P.; Carrell, N.; and McDonagh, J. *J. Cell Biol.* **91**:673–678 (1981).

Ernst, S. A. and Mills, J. W. *J. Cell Biol.* **75**:74–94 (1977).

Esko, J. D.; Nishijima, M.; and Raetz, C. R. H. *Proc. Natl. Acad. Sci.*, U.S. **79**:1698–1702 (1982).

Fambrough, D. M. *Physiol. Revs.* **59**:165–227 (1979).

Fambrough, D. M. and Devreotes, P. N. *J. Cell Biol.* **76**:237–244 (1978).

Fambrough, D. M.; Engel, A. G.; and Rosenberry, T. L. *Proc. Natl. Acad. Sci.*, U.S. **79**:1078–1082 (1982).

Farquhar, M. G. and Palade, G. E. *J. Cell Biol.* **17**:375–412 (1963).

Fawcett, D. W. *The Cell,* 2d ed. W. B. Saunders, Co., Philadelphia, 1981.

Feldman, E. L.; Axelrod, D.; Schwartz, M.; Heacock, A. M.; and Agranoff, B. W. *J. Neurobiol.* **12**:591–598 (1981).

Fernandez, H. L. and Inestrosa, N. C. *Nature,* London **262**:55–56 (1976).

Finean, J. B.; Coleman, R.; and Michell, R. H. *Membranes and their Cellular Functions,* 2d ed. Blackwell, Oxford, 1978.

Finkelstein, A. *Arch. Intern. Med.* **129**:229–242 (1972).

Fischbach, G. C. and Cohen, S. A. *Dev. Biol.* **31**:147–162 (1973).

Fisher, K. A. *J. Cell Biol.* **93**:155–163 (1982).

Fisher, K. A. and Stoeckenius, W. *Science,* **197**:72–74 (1977).

Fitzpatrick-McElligott, S. and Stent, G. S. *J. Neurosci.* **1**:901–907 (1981).

Flagg-Newton, J. L. and Lowenstein, W. R. *Science* **207**:771–773 (1980).

Flagg-Newton, J. L.; Dahl, G.; and Lowenstein, W. R. *J. Memb. Biol.* **63**:105–121 (1981).

Forte, M.; Satow, Y.; Nelson, D.; and Kung, C. *Proc. Natl. Acad. Sci.*, U.S. **78**:7195–7199 (1981).

Francis, D. *Nature,* London **258**:763–765 (1975).

Franks, N. P.; Melchior, V.; Kirschner, D. A.; and Caspar, D. L. *J. Mol. Biol.* **155**:133–153 (1982).

Friedman, D. L. *Physiol Revs.* **56:**652–708 (1976).

Friend, D. S. *J. Cell Biol.* **93:**243–249 (1982).

Friend, D. S. and Gilula, N. B. *J. Cell Biol.* **53:**148–159 (1972).

Fritz, L. C. and Brockes, J. P. *Nature,* London **291:**190–191 (1981).

Froehner, S. C.; Gulbrandsen, V.; Hyman, C.; Jeng, A. Y.; Neubig, R. R.; and Cohen, J. B. *Proc. Natl. Acad. Sci.,* U.S. **78:**5230–5234 (1981).

Frye, L. D. and Edidin, M. *J. Cell Sci.* **7:**319–335 (1970).

Fuchs, S. *Adv. Cytopharm.* **3:**279–285 (1979).

Fukuda, J.; Kameyama, M.; and Yamaguchi, K. *Nature,* London **294:**82–85 (1981).

Fung, B. K-K.; Hurley, J. B.; and Stryer, L. *Proc. Natl. Acad. Sci.,* U.S. **78:**152–156 (1981).

Furshpan, E. J. and Potter, D. D. *Nature,* London **180:**342–343 (1957).

Furshpan, E. J. and Potter, D. D. *J. Physiol.* **145:**289–325 (1959).

Furthmayr, H. *Nature,* London **271:**519–524 (1978).

Furukawa, T. and Furshpan, E. J. *J. Neurophysiol.* **26:**140–176 (1963).

Garavito, R. M. and Rosenbusch, J. P. *J. Cell Biol.* **86:**327–329 (1980).

Garrod, D. R. and Nicol, A. *Biol. Rev.* **56:**199–242 (1981).

Geiger, B. *Trends Biochem. Sci.* **6:**8–9 (1981).

Geiger, B. *Trends Biochem. Sci.* **7:**388–389 (1982).

Geiger, B.; Dutton, A. H.; Tokuyasu, K. T.; and Singer, S. J. *J. Cell Biol.* **91:**614–628 (1981).

Geisow, M. J. *Nature,* London **288:**434–436 (1980).

Gerisch, G.; Fromm, H.; Huesgen, A.; and Wick, U. *Nature,* London **255:**547–549 (1975).

Gilles, R. J. *Trends Biochem. Sci.* **7:**41–42 (1982).

Gilula, N. B. and Satir, P. *J. Cell Biol.* **51:**869–872 (1971).

Gilula, N. B.; Branton, D.; and Satir, P. *Proc. Natl. Acad. Sci.,* U.S. **67:**213–220 (1970).

Gilula, N. B.; Reeves, O. R.; and Steinbach, A. *Nature,* London **235:**262–265 (1972).

Gingell, D. *Mammalian Cell Membranes: General Concepts.* Ed. G. A. Jamieson and D. M. Robinson, Butterworths **1:**198–223 (1976).

Gingle, A. R. *Dev. Biol.* **58:**394–401 (1977).

Gisiger, V.; Vigny, M.; Gautron, J.; and Rieger, F. *J. Neurochem.* **30:**501–516 (1978).

Glauert, A. M. *J. Royal Micros. Soc.* **88**:49–70 (1967).

Glick, M. C. and Flowers, H. *The Glycoconjugates.* Ed. M. Horowitz and W. Pigman, Academic Press, New York **2**:337–384 (1978).

Goldstein, J. L.; Anderson, R. G. W.; and Brown, M. S. *Nature,* London **279**:679–685 (1979).

Gorbsky, G. and Steinberg, M. S. *J. Cell Biol.* **90**:243–248 (1981).

Gorbsky, G.; Cohen, S. M.; and Steinberg, M. S. *J. Cell Biol.* **91**:111a (1981).

Gordon, A. S., and Diamond, I. *J. Supramol. Str.* **14**:163–174 (1980).

Gordon, W. E., III; Bushnell, A.; and Burridge, K. *Cell* **13**:249–261 (1978).

Gorter, E. and Grendel, F. *J. Exp. Med.* **41**:439–443 (1925).

Gozes, I. and Sweadner, K. J. *Nature,* London **294**:477–480 (1981).

Gratzl, M.; Schudt, C.; Ekerdt, R.; and Dahl, G. *Membrane Structure and Function.* Ed. E. E. Bittar, Wiley, New York **3**:59–92 (Chap. 2) (1980).

Greaves, M. F. *Nature,* London **265**:681–683 (1977).

Green, J.; Griffiths, G.; Louvard, D.; Quinn, P.; and Warren, G. *J. Mol. Biol.* **152**:663–698 (1981).

Greengard, P. *Science* **199**:146–152 (1978).

Greengard, P. *Fed. Proc.* **38**:2208–2217 (1979).

Greif, K. F. and Reichardt, L. F. *J. Neurosci.* **2**:843–852 (1982).

Grob, D. (ed.). *Ann. N. Y. Acad. Sci.* **1**–902 (1981).

Gruber, H. and Zenker, W. *Brain Research* **141**:325–334 (1978).

Grumet, M.; Rutishauser, U.; and Edelman, G. M. *Nature,* London **295**:693–695 (1982).

Gundersen, C. B.; Katz, B.; and Miledi, R. *Proc. R. Soc. Lond.* **B 213**:489–493 (1981).

Gutnick, M. J. and Prince, D. A. *Science* **211**:67–70 (1981).

Habermann, E. *Science* **177**:314–322 (1972).

Hagiwara, S. and Byerly, L. *Ann. Rev. Neurosci.* **4**:69–125 (1981).

Hagmann, J. and Fishman, P. H. *J. Biol. Chem.* **255**:2659–2662 (1980).

Hall, Z. W. *J. Neurobiol.* **4**:343–362 (1973).

Hamill, O. P. and Sakmann, B. *Nature,* London **294**:462–464 (1981).

Hanna, R. B.; Reese, T. S.; Ornberg, R. L.; Spray, D. C.; and Bennett, M. V. L. *J. Cell Biol.* **91**:125a (1981).

Hauri, H. P.; Quaroni, A.; and Isselbacher, K. J. *Proc. Natl. Acad. Sci., U.S.* **76**:5183–5186 (1979).

Hayward, S. B. and Stroud, R. M. *J. Mol. Biol.* **151**:491–517 (1981).

Hayward, S. F.; Grano, D. A.; Glaeser, R. M.; and Fisher, K. A. *Proc. Natl. Acad. Sci.,* U.S. **75**:4320–4324 (1978).

Hebdon, G. M. *Neurosci. Res. Prog. Bull.* **20**:321–326 (1982).

Hebdon, G. M.; Le Vine, H., III; Sahyoun, N. E.; Schmitges, C. J.; and Cuatrecasas, P. *Proc. Natl. Acad. Sci.,* U.S. **78**:120–123 (1981).

Hebdon, G. M.; Le Vine, H., III; Sahyoun, N. E.; Schmitges, C. J.; and Cuatrecasas, P. *Biophys. J.* **37**:41–42(1982).

Heidmann, T. and Changeux, J-P. *Ann. Rev. Biochem.* **47**:317–357 (1978).

Henderson, D.; Eibl, H.; and Weber, K. *J. Mol. Biol.* **132**:193–218 (1979).

Henderson, R. *J. Mol. Biol.* **93**:123–138 (1975).

Henderson, R. *Nature,* London **287**:490 (1980).

Henderson, R. and Unwin, P. N. T. *Nature,* London **257**:28–32 (1975).

Henderson, R.; Jubb, J. S.; and Rossmann, M. G. *J. Mol. Biol.* **154**:501–514 (1982).

Henderson, R.; Jubb, J. S.; and Whytock, S. *J. Mol. Biol.* **123**:259–274 (1978).

Henry, N. F. M. and Lonsdale, K. *The International Union of Crystallography,* Vol. 1. The Kynoch Press, Birmingham, England (1965).

Hertzberg, E. L.; Anderson, D. J.; Friedlander, M.; and Gilula, N. B. *J. Cell Biol.* **92**:53–59 (1982).

Herzog, V. *Trends Biochem. Sci.* **6**:319–322 (1981).

Herzog, V. and Farquhar, M. G. *Proc. Natl. Acad. Sci.,* U.S. **74**:5073–5077 (1977).

Heuser, J. E. and Reese, T. S. *J. Cell Biol.* **57**:315–344 (1973).

Hider, R. C. *Nature,* London **292**:803–804 (1981).

Hider, R. C. *Nature,* London **281**:340–341 (1979).

Hirokawa, N. and Heuser, J. *J. Cell Biol.* **91**:122a (1981).

Hoffman, W.; Sarzala, M. G.; Gomez-Fernandez, J. C.; Goni, F. M.; Restall, C. J.; and Chapman, D. *J. Mol. Biol.* **141**:119–132 (1980).

Houslay, M. *Nature,* London **303**:133 (1983).

Howard, F. D.; Petty, H. R.; and McConnell, H. M. *J. Cell Biol.* **92**:283–288 (1982).

Hughes, R. C. *Membrane Glycoproteins.* Buttersworth, London, 1976.

Huxley, A. F. and Stampfli, R. *J. Physiol.* **108**:315–319 (1949).

Ilundain, A. and Naftalin, R. J. *Nature,* London **279**:446–448 (1979).

Inestrosa, N. C.; Reiness, C. G.; Reichardt, L. F.; and Hall, Z. W. *J. Neurosci.* **1:**1260–1267 (1981).

Ingolia, T. D. and Koshland, D. E., Jr. *J. Biol. Chem.* **253:**3821–3829 (1978).

Isräel, M.; Lesbats, B.; and Manaranche, R. *Nature,* London **294:**474–475 (1981).

Ito, S. *J. Cell Biol.* **27:**475–491 (1965).

Jackson, M. B.; Stephens, C. L.; and Lecar, H. *Proc. Natl. Acad. Sci., U.S.* **78:**6421–6425 (1981).

Jacobs, S. and Cuatrecasas, P. *Trends Biochem. Sci.* **2:**280–282 (1977).

Jacobson, K.; Elson, E.; Koppel, D.; and Webb, W. *Nature,* London **295:**283–284 (1982).

Jakobs, K. H.; Aktories, K.; and Schultz, G. *Nature,* London **303:**177–178 (1983).

Jap, B. K.; Maestre, M. F.; Hayward, S. B.; and Glaeser, R. M. *Biophys. J.* (in press).

Jennings, K. R.; Host, J. J.; Kaczmarek, L. K.; and Strumwasser, F. *J. Neurobiol.* **12:**579–590 (1981).

Jessell, T. M.; Siegel, R. E.; and Fishbach, G. D. *Proc. Natl. Acad. Sci., U.S.* **76:**5397–5401 (1979).

Jones, G. E. and Witkowski, J. A. *J. Cell Sci.* **48:**291–300 (1981).

Kachar, B. and Reese, T. S. *Nature,* London **296:**464–466 (1982).

Kaczmarek, L. K.; Jennings, K. R.; Strumwasser, F.; Nairn, A. C.; Walter, U.; Wilson, F. D.; and Greengard, P. *Proc. Natl. Acad. Sci., U.S.* **77:**7487–7491 (1980).

Kalcheim, C.; Vogel, Z.; and Duksin, C. *Proc. Natl. Acad. Sci., U.S.* **79:**3077–3081 (1982).

Kaplan, J. *Science* **212:**14–20 (1981).

Karlin, A.; Damle, V.; Hamilton, S.; McLaughlin, M.; Valderrama, R.; and Wise, D. *Adv. Cytopharm.* **3:**183–189 (1979).

Karnovsky, M. J.; Kleinfeld, A. M.; Hoover, R. L.; and Klausner, R. D. *J. Cell Biol.* **94:**1–6 (1982).

Katz, B. and Miledi, R. *Proc. Roy. Soc. Lond.* **B 167:**23–38 (1967).

Kehry, M.; Yguerabide, J.; and Singer, S. J. *Science* **195:**486–487 (1977).

Kell, D. B. *Trends Biochem. Sci.* **6:**8–9 (1981).

Kelly, R. B.; Deutch, J. W.; Carlson, S. S.; and Wagner, J. A. *Ann. Rev. Neurosci.* **2:**399–446 (1979).

Kennedy, M. B. and Greengard, P. *Proc. Natl. Acad. Sci., U.S.* **78**:1293–1297 (1981).

Kennedy, S. J. *J. Memb. Biol.* **42**:265–279 (1978).

Khorana, H.; Gerber, G.: Herlihy, W.; Gray, C.; Anderegg, R.; Nihei, K.; and Biemann, K. *Proc. Natl. Acad. Sci., U.S.* **76**:5046–5050 (1979).

Kilbourn, B. T.; Dunitz, J. D.; Pioda, L. A.; and Simon, J. *J. Mol. Biol.* **30**:559–563 (1967).

Kimura, Y.; Ikegami, A.; Ohno, K.; Takeuchi, Y.; and Saigo, S. *Photochem. Photobiol.* **33**:435–439 (1981).

King, G. I.; Stoeckenius, W.; Crepsi, H.; and Schoeborn, B. *J. Mol. Biol.* **130**:395–404 (1979).

King, G.; Stoeckenius, W.; Crespi, H. L.; and Schoenborn, B. P. *Proc. Natl. Acad. Sci., U.S.* **77**:4726–4730 (1980).

Kirschner, D. A. and Ganser, A. L. *Nature,* London **283**:207–210 (1980).

Kistler, J. and Bullivant, S. *J. Ultrastructure Res.* **72**:27–38 (1980).

Kistler, J.; Stroud, R. M.; Klymkowsky, W.; Lalancette, A.; and Fairclough, R. H. *Biophys. J.* **37**:371–383 (1982).

Kleeman, W. and McMonnell, H. M. *Biochim. Biophys. Acta.* **345**:220–230 (1974).

Klein, M. and Kandel, E. R. *Proc. Natl. Acad. Sci., U.S.* **75**:3512–3516 (1978).

Klymkowsky, M. W. and Stroud, R. M. *J. Mol. Biol.* **128**:319–334 (1979).

Koenig, J. and Vigny, M. *Nature,* London **271**:75–77 (1978).

Koeppe, R. E., II; Hodgson, K. D.; and Stryer, L. *J. Mol. Biol.* **121**:41–54 (1978).

Köhler, G. and Milstein, C. *Nature,* London **256**:495–497 (1975).

Konijn, T. M.; Barkley, D. S.; Chang, Y. Y.; and Bonner, J. T. *Am. Nat.* **102**:225–233 (1968).

Kouyama, T.; Kimura, Y.; Kinosita, K., Jr.; and Ikegami, A. *J. Mol. Biol.* **153**:337–359 (1981).

Kracke, G. R.; O'Neal, S. G.; and Chacko, G. K. *J. Memn. Biol.* **63**:147–156 (1981).

Kretsinger, R. H. *Neurosci. Res. Prog. Bull.* **19**:217–328 (1981).

Kristol, C.; Sandri, C.; and Akert, K. *Brain Res.* **142**:391–400 (1978).

Kuffler, S. W. and Nicholls, J. G. *From Neuron to Brain.* Sinauer Associates, Inc., Sunderland, Mass., 1976.

Kupfer, A.; Louvard, D.; and Singer, S. J. *Proc. Natl. Acad. Sci., U.S.* **79:**2603–2607 (1982).

Kyte, J. and Doolittle, R. F. *J. Mol. Biol.* **157:**105–132 (1982).

Lackie, J. M. *Membrane Structure and Function.* Ed. E. E. Bittar, Wiley, New York **1:**73–102 (1980).

Lanyi, J. K. *Trends Biochem. Sci.* **6:**60 (1981).

Lasansky, A. *J. Cell Biol.* **40:**577–581 (1969).

Lasek, R. J. *Neurosci. Res. Prog. Bull.* **19:**7–32 (1981).

Lau, A. L. Y. and Chan, S. I. *Proc. Natl. Acad. Sci., U.S.* **72:**2170–2174 (1975).

Lawrence, J. C. and Catterall, W. A. *J. Biol. Chem.* **256:**6213–6222 (1981).

Lawrence, T. S.; Beers, W. H.; and Gilula, N. B. *Nature,* London **272:**501–506 (1978).

Lawson, D.; Raff, M. C.: Gomperts, B.; Fewtrell, C.; and Gilula, N. B. *J. Cell Biol.* **72:**242–259 (1977).

Lazarides, E. *Nature,* London **283:**249–256 (1980).

Lee, C. Y. *Adv. Cytopharm.* **3:**1–16 (1979).

Lester, H. A. *Nature,* London **294:**398–399 (1981).

Lewin, R. *Science* **211:**38–42 (1981).

Lim, V. *Protein Folding.* Ed. R. Jaenicke, Elsveier, Amsterdam, 1980, pp. 149–166.

Lin, K.; Weis, R. M.; and McConnell, H. M. *Nature,* London **296:**164–165 (1982).

Lindstrom, J. *Adv. Cytopharm.* **3:**245–253 (1979).

Lindstrom, J.; Einarson, B.; and Merlie, J. *Proc. Natl. Acad. Sci., U.S.* **75:**769–773 (1978).

Lodish, H. F.; Braell, W. A.; Schwartz, A. L.; Strous, G.; and Zilberstein, A. *Int. Rev. Cytology,* Suppl. 12. Ed. A. L. Muggleton-Harris, Academic Press, New York, 1981, pp. 247–307.

Loewenstein, W. R. *Physiol. Revs.* **61:**829–913 (1981).

Longmuir, K. J.; Capaldi, R. A.; and Dahlquist, F. W. *Biochem.* **16:**5746–5755 (1977).

Lucy, J. A. *Cell Membranes.* Ed. G. Weissmann, and R. Claiborne, Hospital Practice Publishing Co., New York, 1975, pp. 75–83.

Lux, S. E. *Nature,* London **281:**426–429 (1979).

Luzzati, V. and Husson, F. *J. Cell Biol.* **12:**207–219 (1962).

MacGregor, C. H. and McElhaney, G. E. *J. Bacteriology* **148**:551–558 (1981).

Madrid, R. E.; Jaros, E.; Cullen, M. J.; and Bradley, W. G. *Nature,* London **257**:319–321 (1975).

Makowski, L.; Caspar, D. L. D.; Goodenough, D. A.; and Philips, W. C. *Biophys. J.* **37**:189–191 (1982).

Makowski, L.; Caspar, D. L. D.; Phillips, W. C.; and Goodenough, D. A. *J. Cell Biol.* **4**:629–645 (1977).

Malhotra, S. K. *J. Ultrastructure Res.* **15**:14–37 (1966).

Malhotra, S. K. *Subcell. Biochem.* **8**:273–309 (1981).

Malhotra, S. K. *Membrane Structure and Function.* Ed. E. E. Bittar, Wiley, New York, **1**:1–72 (1980).

Malhotra, S. K. *Mammalian Cell Membranes.* Ed. G. A. Jamieson and D. M. Robinson, Butterworths, London, **1**:224–243 (1976).

Malhotra, S. K. and Tewari, J. *Cytobios* **34**:83–96 (1982).

Malhotra, S. K. and Tipnis, U. R. *Proc. R. Soc. Lond.* **B 203**:59–68 (1978).

Malhotra, S. K.; Ross, S.; and Tewari, J. P. *Chem. Phys. Lipids* **28**:33–39 (1981).

Malhotra, S. K.; Tewari, J. P.; and Tu, J. C. *J. Neurobiol.* **6**:57–71 (1975).

Manjunath, C. K.; Goings, G. E.; and Page, E. *J. Cell Biol.* **91**:100a (1981).

Marchesi, V. T. and Steers, E., Jr. *Science* **159**:203–204 (1968).

Marchesi, V. T.; Jackson, R. L.; Segrest, J. P.; and Kahane, I. *Fed. Proc.* **32**:1833–1837 (1973).

Marinetti, G. V. and Cattieu, K. *J. Biol. Chem.* **257**:245–248 (1982).

Marinetti, G. V. and Crain, R. C. *J. Supramol. Str.* **8**:191–213 (1978).

Markham, R.; Frey, S.; and Hills, G. J. *Virology* **20**:88–102 (1963).

Marnay, A. and Nachmansohn, D. *J. Physiol.* **92**:37–47 (1938).

Marshall, L. M. *Proc. Natl. Acad. Sci., U.S.* **78**:1948–1952 (1981).

Mason, W. T.; Fager, R. S.; and Abrahamson, E. W. *Nature,* London **247**:188–191 (1974).

Massoulié, J. *Trends Biochem. Sci.* **5**:160–164 (1980).

Matthieu, J. M.; Market, M.; Vanier, M. T.; Rutti, M.: Reigner, J.; and Bourre, J. M. *Brain Res.* **226**:235–244 (1981).

McMahan, U. J.; Sanes, J. R.; and Marshall, L. M. *Nature,* London **271**:172–174 (1978).

McManaman, J. L.; Blosser, J. C.; and Appel, S. H. *J. Neurosci.* **1**:771–776 (1981).

Merlie, J. P. and Sebbane, R. *J. Biol. Chem.* **256**:3605–3608 (1981).

Merlie, J. P.; Hofler, J. G.; and Sebbane, R. *J. Biol. Chem.* **256**:6995–6999 (1981).

Mescher, M. F.; Jose, M. J. L.; and Balk, S. P. *Nature,* London **289**:139–144 (1981).

Metzger, H. and McConnell, H. M. *Neurosci. Res. Prog. Bull.* **20**:355–370 (1982).

Michel, H.; Oesterhelt, D.; and Henderson, R. *Proc. Natl. Acad. Sci., U.S.* **77**:338–342 (1980).

Michell, R. H. *Neurosci. Res. Prog. Bull.* **20**:338–350 (1982).

Miledi, R. *Proc. R. Soc. Lond.* **B 183**:421–425 (1973).

Miledi, R. *J. Physiol.* **151**:24–30 (1960).

Miledi, R. and Uchitel, O. D. *Proc. R. Soc. Lond.* **B 213**:243–248 (1981).

Miller, K. R. and Lassignal, N. L. *J. Cell Biol.* **91**:304a (1981).

Milstein, C. *Proc. R. Soc. Lond.* **B 211**:393–412 (1981).

Milstein, C.; Brownlee, G. G.; Harrison, T. M.; and Mathews, M. B. *Nature,* London **239**:117–120 (1972).

Mitchell, P. *Biol. Revs.* Camb. Philosophical Soc., Cambridge, England, **41**:445–502 (1966).

Mochly-Rosen, D. and Fuchs, S. *Biochemistry* **20**:5920–5924 (1981).

Møllgard, K.; Lauiritzen, B.; and Saunders, N. R. *J. Neurocytology* **8**:139–149 (1979).

Montal, M.; Darszon, A., and Schindler, H. *Quarterly Reviews Biophys.* **14**:1–79 (1981).

Montesano, R.; Perrelet, A.; Vassalli, P.; and Orci, L. *Proc. Natl. Acad. Sci., U.S.* **76**:6391–6395 (1979).

Moore, H. P.; Fritz, L. C.; Raftery, M. A.; and Brockes, J. P. *Proc. Natl. Acad. Sci., U.S.* **79**:1673–1677 (1982).

Morel, M.; Manaranche, R.; Israel, M.; and Gulik-Krzywicki, T. *J. Cell Biol.* **93**:349–356 (1982).

Morley, B. J. and Kemp, G. E. *Brain Res. Revs.* **3**:81–104 (1981).

Mueller, P.; Rudin, D. O.; Tien, H. T.; and Wescott, W. C. *Nature,* London **194**:979–980 (1962).

Murthy, S. N. P.; Liu, T.; Kaul, R. K.; Kohler, H.; and Steck, T. L. *J. Biol. Chem.* **256:**11203–11208 (1981).

Nachmansohn D. *The Harvey Lectures.* Academic Press, New York, 1955, pp. 57–99.

Nägeli, C. *Circulation* **26:**987–1012 (1855). Quoted in Smith, 1962.

Nagle, J. F. and Morowitz, H. J. *Proc. Natl. Acad. Sci., U.S.* **75:**298–302 (1978).

Narahashi, T. *Physiol. Rev.* **54:**813–819 (1974).

Nathanson, N. M. and Hall, Z. W. *Biochem.* **18:**3392–3401 (1979).

Neher, E. and Sakmann, B. *Nature,* London **260:**799–802 (1976).

Neidle, S.; Berman, H. M.; and Shieh, H. S. *Nature,* London **288:**129–133 (1980).

Nelson, G.; Roberts, T.; and Ward, S. *J. Cell Biol.* **92:**121–131 (1982).

Nelson, N.; Anholt, R.; Linstrom, J.; and Montal, M. *Proc. Natl. Acad. Sci., U.S.* **77:**3057–3061 (1980).

Nermut, M. V. *Europ. J. Cell Biol.* **25:**265–271 (1981).

Newell, P. C. *Endeavour, New Series* **1:**63–68 (1977).

Nicholson, B. J.; Hunkapiller, M. W.; Hood, L. E.; Revel, J. P.; and Takemoto, L. *J. Cell Biol.* **87:**200a (1980).

Nielsen, T. B.; Lad, P. M.; Preston, M. S.; Kempner, E.; Schlegel, W.; and Rodbell, M. *Proc. Natl. Acad. Sci., U.S.* **78:**722–726 (1981).

Nieto-Sampedro, M.; Bussineau, C. M.; and Cotman, C. W. *J. Neurosci.* **2:**722–734 (1982).

Noda, M.; Takahashi, H.; Tanabe, T.; Toyosato, M.; Furutani, Y.; Hirose, T.; Asai, M.; Inoyama, S.; Miyata, T.; and Numa, S. *Nature,* London **299:**793–797 (1982).

Norman, R. I.: Mehraban, F.: Barnard, E. A.; and Dolly, J. O. *Proc. Natl. Acad. Sci., U.S.* **79:**1321–1325 (1982).

Oesterhelt, D. and Stoeckenius, W. *Nature,* London **233:**149–152 (1971).

Op den Kamp, J. A. F. *Ann. Rev. Biochem.* **48:**47–71 (1979).

Orci, L.; Perrelet, A.; and Friend, D. S. *J. Cell Biol.* **75:**23–30 (1977).

Orci, L.; Singh, A.; Amherdt, M.; Brown, D.; and Perrelet, A. *Nature,* London **293:**646–647 (1981).

Orly, J. and Schramm, M. *Proc. Natl. Acad. Sci., U.S.* **73:**4410–4414 (1976).

Ornberg, R. L. and Reese, T. S. *J. Cell Biol.* **90:**40–54 (1981).

Oschman, J. L. *Membrane Structure and Function.* Ed. E. E. Bittar, Wiley, New York **2**:141–170 (1980).

Oswald, R. E. and Freeman, J. A. *J. Biol. Chem.* **254**:3419–3426 (1979).

Ovchinnikov, Y. A.; Abdulaev, N. G.; Feigina, M. Y.; Kiselev, A. V.; and Lobanov, N. A. *FEBS Letts.* **100**:219–224 (1979).

Paiement, J.; Beaufay, H.; and Godelaine, D. *J. Cell. Biol.* **86**:29–37 (1980).

Papahadjopoulos, D.; Portis, A.; and Pangborn, W. *Annals N. Y. Acad. Sci.* **308**:50–66 (1978).

Pasvol, G.; Wainscoat, J. S.; and Weatherall, D. J. *Nature,* London **297**:64–66 (1982).

Patrick, J.; Stallcup, W. P.; Zavanelli, M.; and Ravdin, P. *J. Biol. Chem.* **255**:526–533 (1980).

Pearse, B. M. F. *Trends Biochem. Sci.* **5**:131–134 (1980).

Pearse, B. M. F. *J. Mol. Biol.* **97**:93–98 (1975).

Pearse, B. M. F. *Proc. Natl. Acad. Sci.,* U.S. **79**:451–455 (1982).

Pearse, B. M. F. *Proc. Natl. Acad. Sci.,* U.S. **73**:1255–1259 (1976).

Peng, H. B.; Cheng, P. C.; and Luther, P. W. *Nature,* London **292**:831–834 (1981).

Pentreath, V. W. and Kai-Kai, M. A. *Nature,* London **295**:59–61 (1982).

Peracchia, C. *Int. Rev. Cytol.* **66**:81–146 (1980).

Peracchia, C. and Peracchia, L. L. *J. Cell Biol.* **87**:708–718 (1980a).

Peracchia, C. and Peracchia, L. L. *J. Cell Biol.* **87**:719–727 (1980b).

Peracchia, C.; Bernardini, G.; and Peracchia, L. L. *J. Cell Biol.* **91**:124a (1981).

Perkins, M. *J. Cell Biol.* **90**:563–567 (1981).

Perrelet, A.; Garcia-Segura, L-M.; Singh, A.; and Orci, L. *Proc. Natl. Acad. Sci.,* U.S. **79**:2598–2602 (1982).

Pestronk, A. and Drachman, D. B. *Nature,* London **288**:733–734 (1980a).

Pestronk, A. and Drachman, D. B. *Science* **210**:342–343 (1980b).

Pfeiffer, J. R.; Oliver, J. M.; and Berlin, R. D. *Nature,* London **286**:727–729 (1980).

Pfenninger, K. H. *Neurosci. Res. Prog. Bull.* **20**:73–79 (1981).

Pierschbacher, M. D.; Hayman, E. G.; and Ruoslahti, E. *Cell* **26**:259–267 (1981).

Pilwat, G.; Richter, H-P.; and Zimmermann, U. *FEBS Letts.* **133**:169–174 (1981).

Pimplikar, S. W. and Malhotra, S. K. *FEBS Letts.* in press (1983).

Pinder, J. C.; Phethean, J.; and Gratzer, W. B. *FEBS Letts.* **92:**278–282 (1978).

Pinto da Silva, P. and Torrisi, M. R. *J. Cell Biol.* **93:**463–469 (1982).

Pinto da Silva, P.; Parkinson, C.; and Dwyer, N. *J. Histochem.* **29:**917–928 (1981).

Podleski, T. R.; Axelrod, D.; Ravdin, P.; Greenberg, I.; Johnson, M. M.; and Saltpeter, M. M. *Proc. Natl. Acad. Sci.*, U.S. **75:**2035–2039 (1978).

Poo, M. M. *Nature,* London **295:**332–334 (1982).

Popot, J. L.; Cartaud, J.; and Changeux, J-P. *Europ. J. Biochem.* **118:**203–214 (1981).

Poste, G. and Allison, A. C. *Biochim. Biophys. Acta* **300:**421–465 (1973).

Privat, A.; Jacque, C.; Bourre, J. M.; Dupouey, P.; and Baumann, N. *Neurosci. Lett.* **12:**107–112 (1979).

Prives, J. and Shinitzky, M. *Nature,* London **268:**761–763 (1977).

Prives, J.; Fulton, A. B.; Penman, S.; Daniels, M. P.; and Christian, C. N. *J. Cell Biol.* **92:**231–236 (1982).

Pumplin, D. W. and Fambrough, D. M. *Ann. Rev. Physiol.* **44:**319–335 (1982).

Pumplin, D. W.; Reese, T. S.; and Llinas, R. *Proc. Natl. Acad. Sci.*, U.S. **78:**7210–7213 (1981).

Racker, E. *A New Look at Mechanisms in Bioenergetics.* Academic Press, New York, 1976.

Raff, M. C. and de Petris, S. *Fed. Proc.* **32:**48–54 (1973).

Raftery, M. A.; Hunkapiller, M. W.; Strader, C. D.; and Hood, L. E. *Science* **208:**1454–1457 (1980).

Rajaraman, R.; Macsween, J. M.; and Fox, R. A. *J. Theor. Biol.* **74:**177–201 (1978).

Rakic, P. and Goldman-Rakic, P. S. *Neurosci. Res. Prog. Bull.* **20:**433–606 (1982).

Ramachandran, G. N. and Sasisekharan, V. *Adv. Prot. Chem.* Academic Press, New York **23:**283–437 (1968).

Rambourg, A. and Leblond, C. P. *J. Cell Biol.* **32:**27–53 (1967).

Ramm, L. E.; Whitlow, M. B.; and Mayer, M. M. *Proc. Natl. Acad. Sci.*, U.S. **79:**4751–4755 (1982).

Rash, J. E.; Hudson, C. S.; and Ellisman, M. H. *Cell Membrane Receptors for Drugs and Hormones: A Multidisciplinary Approach.* Ed. I. Bolis and R. W. Straud, Raven Press, New York, 1978, pp. 47–68.

Razi-Naqvi, K.; Gonzalez-Rodriguez, J.; Cherry, R. J.; and Chapman, D. *Nature New Biol.* **245**:249–251 (1973).

Reaven, E. and Azhar, S. *J. Cell Biol.* **89**:300–308 (1981).

Rehm, R. and Betz, H. *Biochem. Biophys. Res. Comm.* **102**:1385–1392 (1981).

Reuter, H. *J. Physiol.* **242**:429–451 (1974).

Revel, J. P. and Karnovsky, M. J. *J. Cell Biol.* **33**:c7–c12 (1967).

Richman, D. P. and Arnason, B. *Proc. Natl. Acad. Sci.*, U.S. **76**:4632–4635 (1979).

Ritchie, J. M. and Rogart, R. B. *Proc. Natl. Acad. Sci.*, U.S. **74**:211–215 (1977).

Robenek, H.; Jung, W.; and Gelhardt, R. *J. Ultrastr. Res.* **78**:95–106 (1982).

Roberts, T. M. and Ward, S. *J. Cell Biol.* **92**:113–120 (1982a).

Roberts, T. M. and Ward, S. *J. Cell Biol.* **92**:132–138 (1982b).

Robertson, A. D. J. and Grutsch, J. F. *Cell* **24**:603–611 (1981).

Robertson, A. D. J.; Grutsch, J. F.; and Gingle, A. R. *Science* **199**:990–991 (1978).

Robertson, J. D. *J. Cell Biol.* **19**:201–221 (1963).

Robertson, J. D. *Prog. Biophys.* **10**:343–418 (1960).

Robertson, J. D. *Demyelinating Disease: Basic and Clinical Electrophysiology.* Ed. S. G. Waxman and J. M. Ritchie, Raven Press, New York, 1981, pp. 419–477.

Robertson, J. D. and Vergara, J. *J. Cell Biol.* **86**:514–528 (1980).

Robertson, J. D.; Schreil, W.; and Reedy, M. *38th Ann. Proc. E. M. Soc. America,* 1980, p. 792.

Robnek, H.; Jung, W.; and Gebhardt, R. *J. Ultrastr. Res.* **78**:95–106 (1982).

Rodbell, M. *Nature,* London **284**:17–21 (1980).

Rodewald, R. *J. Cell Biol.* **58**:189–211 (1973).

Rodriguez-Boulan, E. and Sabatini, D. *Proc. Natl. Acad. Sci.*, U.S. **75**:5071–5075 (1978).

Rogan, P. K. and Zaccai, G. *J. Mol. Biol.* **145**:281–284 (1981).

Rose, G. D. *Nature,* London **272**:586–590 (1978).

Rose, G. D. and Roy, S. *Proc. Natl. Acad. Sci.*, U.S. **77**:4643–4647 (1980).

Rosenmann, E.; Kreis, C.; Thompson, R. G.; Dobbs, M.; Hamerton, J. L.; and Wrogemann, K. *Nature,* London **298**:563–565 (1982).

Ross, M. J.; Klymkowsky, M. W.; Agard, D. A.; and Stroud, R. M. *J. Mol. Biol.* **116**:635–659 (1977).

Rossmann, M. G. (ed.). *The Molecular Replacement Method.* Gordon and Breach, New York, 1972.

Rossmann, M. G. and Henderson, R. *Acta Cryst.* **A38**:13–20 (1982).

Roth, S. *Quart. Rev. Biol.* **48**:541–563 (1973).

Roth, T. F. and Porter, K. R. *J. Cell Biol.* **20**:313–332 (1964).

Rothman, J. E.; Katz, F. N.; and Lodish, H. F. *Cell* **15**:1447–1454 (1978).

Rottem, S. *FEBS Letts.* **133**:161–164 (1981).

Rotundo, R. L. and Fambrough, D. M. *J. Biol. Chem.* **254**:4790–4799 (1979).

Sabatini, D. D.; Kriebich, G.; Morimoto, T.; and Adesnik, M. *J. Cell Biol.* **92**:1–22 (1982).

Sahyoun, N. E.; Le Vine, H., III; Hebdon, G. M.; Hemadah, R.; and Cuatrecasas, P. *Proc. Natl. Acad. Sci.*, U.S. **78**:2359–2362 (1981).

Saibil, H. *Nature,* London **297**:106–108 (1982).

Sakakibara, K.; Momoi, T.; Uchida, T.; and Nagai, Y. *Nature,* London **293**:76–78 (1981).

Salhany, J. M. and Gaines, K. C. *Trends Biochem. Sci.* **6**:13–15 (1981).

Salisbury, J. L.; Condeelis, J. S.; Maihle, N. J.; and Satir, P. *Nature,* London **294**:163–166 (1981).

Sandermann, H., Jr. *Biochim. Biophys. Acta.* **515**:209–237 (1978).

Scandella, C.; Campisi, J.; Elhai, J.; and Selak, M. *Biophys. J.* **37**:16–17 (1982).

Schindler, H. and Rosenbusch, J. P. *Proc. Natl. Acad. Sci.*, U.S. **78**:2302–2306 (1981).

Schindler, M. and Rosenbusch, J. P. *J. Cell Biol.* **92**:742–746 (1982).

Schlessinger, J. *Trends Biochem. Sci.* **5**:210–214 (1980).

Schlessinger, J.; Axelrod, D.; Koppel, D. E.; Webb, W. W.; and Elson, E. L. *Science* **195**:307–309 (1977).

Schmidt, M. F. G. and Schlesinger, M. J. *Cell* **17**:813–819 (1979).

Schmitt, F. O. and Samson, F. E., Jr. *Neurosci. Res. Prog. Bull.* **7**:281–417 (1969).·

Schneider, Y-J.; de Duve, C.; and Trouet, A. *J. Cell Biol.* **88**:380–387 (1981).

Schofield, G. G.; Witkop, B.; Warnick, J. E.; and Albuquerque, E. X. *Proc. Natl. Acad. Sci.*, U.S. **78**:5240–5244 (1981).

Schook, W.; Puszkin, S.; Bloom, W.; Ores, C.; and Kochwa, S. *Proc. Natl. Acad. Sci.*, U.S. **76**:116–120 (1979).

Schulman, H.; Hunter, W. B.; and Greengard, P. *Calcium and Cell Function.* Ed. W. Y. Cheung, Academic Press, New York **1**:220–252 (Chap. 11) (1980).

Schwabe, U.; Puchstein, C.; Hannemann, H.; and Sochtig, E. *Nature, London* **277**:143–145 (1979).

Sealock, R. *J. Cell Biol.* **92**:514–522 (1982).

Seite, R.; Leonetti, J.; Luciani-Vuillet, J.; and Bio, M. *Brain Res.* **124**:41–51 (1977).

Sen, A.; Williams, W. P.; Brain, A. P. R.; Dickens, M. J.; and Quinn, P. J. *Nature, London* **293**:488–490 (1981).

Shaffer, B. M. *Nature, London* **255**:549–552 (1975).

Sheridan, J. D. *J. Cell Biol.* **37**:650–659 (1968).

Sheridan, J. D.; Hammer-Wilson, M.; Preus, D.; and Johnson, R. G. *J. Cell Biol.* **76**:532–544 (1978).

Sherman, J. M.; Sabatini, D. D.; and Morimoto, T. *J. Cell Biol.* **87**:307a (1980).

Shotton, D. M.; Burke, B. E.; and Branton, D. *J. Mol. Biol.* **131**:303–329 (1979).

Sikerwar, S. and Malhotra, S. K. *Europ. J. Cell Biol.* **25**:319–323 (1981).

Sikerwar, S. S.; Tewari, J. P.; and Malhotra, S. K. *Europ. J. Cell Biol.* **24**:211–215 (1981).

Silberstein, L.; Inestrosa, N. C.; and Hall, Z. W. *Nature, London* **295**:143–145 (1982).

Silverstein, S. C.; Steinman, R. M.; and Cohn, Z. A. *Ann. Rev. Biochem.* **46**:669–772 (1977).

Simionescu, N.; Simionescu, M., and Palade, G. E. *J. Cell Biol.* **90**:605–613 (1981a).

Simionescu, M.; Simionescu, N. M.; Silbert, J. E.; and Palade, G. E. *J. Cell Biol.* **90**:614–621 (1981b).

Simons, K.; Garoff, H.; and Helenius, A. *Sci. Amer.* **246**:58–66 (1982).

Sinensky, M.; Pinkerton, F.; Sutherland, E.; and Simon, F. R. *Proc. Natl. Acad. Sci.*, U.S. **76**:4893–4897 (1979).

Singer, S. J. and Nicolson, G. L. *Science* **175:**720–731 (1972).

Small, J. V. *J. Cell Biol.* **91:**695–705 (1981).

Smilowitz, H.; Hadjian, R. A.; Dwyer, J.; and Feinstein, M. B. *Proc. Natl. Acad. Sci.,* U.S. **78:**4708–4712 (1981).

Smith, D. K. and Palek, J. *Nature,* London **297:**424–425 (1982).

Snider, M. D. *Nature,* London **298:**117–118 (1982).

Spray, D. C.; Harris, A. L.; and Bennett, M. V. L. *Science* **211:**712–715 (1981a).

Spray, D. C.; Harris, A. L.; and Bennett, M. V. L. *Biophys. J.* **33:**108 A (1981b).

St. John, P. A.; Froehner, S. C.; Goodenough, D. A.; and Cohen, J. B. *J. Cell Biol.* **92:**333–342 (1982).

Stahl, P. D. and Schlesinger, P. H. *Trends Biochem. Sci.* **5:**194–200 (1980).

Steere, R. L. *J. Biophys. Biochem. Cytol.* **3:**45–60 (1957).

Steinhardt, R. A. and Alderton, J. M. *Nature,* London **295:**154–155 (1982).

Steinhardt, R. A. and Epel, D. *Proc. Natl. Acad. Sci.,* U.S. **71:**1915–1919 (1974).

Stevens, C. F. *Nature,* London **287:**13–14 (1980).

Stevens, C. F. *Nature,* London **299:**776–777 (1982).

Stoeckenius, W. *Sci. Amer.* **234:**38–46 (1976).

Stoeckenius, W. *Accounts of Chemical Research* **13:**337–344 (1980).

Stoeckenius, W. *Circulation* **26:**1066–1069 (1962).

Strumwasser, F.; Kaczmarek, L. K.; Jennings, K. R.; and Chiu, A. Y. *Eighth Internat. Symposium On Neurosecretion.* Friday Harbor Labs, Friday Harbor, Washington 1980, pp. 4–10.

Stryer, L. *Biochemistry,* 2d ed. Freeman and Co., 1981.

Stryer, L.; Hurley, J. B.; and Fung, B. K. *Trends Biochem. Sci.* **6:**245–247 (1981).

Stühmer, W. and Almers, W. *Proc. Natl. Acad. Sci.,* U.S. **79:**946–950 (1982).

Stya, M. and Axelrod, D. *J. Cell Biol.* **91:**121a (1981).

Stya, M. and Axelrod, D. *Proc. Natl. Acad. Sci.,* U.S. **80:**449–453 (1983).

Sutherland, E. W. *Science* **177:**401–408 (1972).

Tamm, S. L. *J. Cell Biol.* **80:**141–149 (1979).

Terwilliger, T. C.; Weissman, L.; and Eisenberg, D. *Biophys. J.* **37:**353–361 (1982).

Tewari, J. P.; Sehgal, S. S.; and Malhotra, S. K. *J. Histochem. Cytochem.* **30**:436–440 (1982).

Thiery, J. P.; Brackenbury, R.; Rutishauser, U.; and Edelman, G. M. *J. Biol. Chem.* **252**:6841–6845 (1977).

Tillack, T. W. and Marchesi, V. T. *J. Cell Biol.* **45**:649–653 (1970).

Tilney, L. G. *International Cell Biol.* 1976/1977, pp. 388–402.

Tipnis, U. and Malhotra, S. K. *FEBS Lett.* **69**:141–143 (1976).

Tipnis, U. R. and Malhotra, S. K. *Canad. J. Physiol. Pharmacol.* **58**:445–458 (1980).

Tipnis, U. R. and Malhotra, S. K. *J. Supramol. Str.* **12**:321–334 (1979).

Tipnis, U. R. and Malhotra, S. K. *Cytobios.* **31**:91–106 (1981).

Tomita, M. and Marchesi, V. T. *Proc. Natl. Acad. Sci.*, U.S. **72**:2964–2968 (1975).

Tosteson, D. C. *Sci. Amer.* **244**:164–174 (1981).

Tschopp, J.; Muller-Eberhard, H. J.; and Podack, E. R. *Nature,* London **298**:534 (1982).

Tu, J. C. and Malhotra, S. K. *Canadian J. Microbiol.* **23**:378–388 (1977).

Tu, J. C. and Malhotra, S. K. *J. Histochem. Cytochem.* **21**:1041–1046 (1973).

Tu, J. C. and Malhotra, S. K. *Cytobios.* **13**:217–228 (1975).

Turin, L. and Warner, A. *Nature,* London **270**:56–57 (1977).

Tweto, J. and Doyle, D. *The Synthesis, Assembly and Turnover of Cell Surface Components.* Ed. G. Poste and G. L. Nicolson, Elsevier, 1977, pp. 137–163.

Ulbricht, W. *Physiol. Revs.* **61**:785–828 (1981).

Unwin, P. N. T. and Zampighi, G. *Nature,* London **283**:545–549 (1980).

Van Deurs, B. and Koehler, J. K. *J. Cell Biol.* **80**:662–673 (1979).

Van Harreveld, A.; Crowell, J.; and Malhotra, S. K. *J. Cell Biol.* **25**:117–137 (1965).

Van Heyningen, S. *Nature,* London **292**:293–294 (1981).

Verkleij, A.; Mombers, C.; Leunissen-Bijvelt, J.; and Ververgaert, P. *Nature,* London **279**:162–163 (1979).

Von Wedel, R. J.; Carlson, S. S.; and Kelly, R. B. *Proc. Natl. Acad. Sci.*, U.S. **78**:1014–1018 (1981).

Walter, P. and Blobel, G. *J. Cell Biol.* **91**:551–556 (1981).

Warner, A. E. *J. Physiol.*, London **235**:267–286 (1973).

Warner, A. E. and Lawrence, P. A. *Cell* **28**:243–252 (1982).

Warren, G. B.; Houslay, M. D.; and Metcalfe, J. C. *Nature,* London **255**:684–687 (1975).

Watts, A. *Nature,* London **294**:512–513 (1981).

Waxman, S. G. and Foster, R. E. *Proc. R. Soc. Lond.* **B 209**:441–446 (1980a).

Waxman, S. G. and Foster, R. E. *Brain Res. Revs.* **2**:205–234 (1980b).

Weatherbee, J. A. *Int. Rev. Cytology,* Supp. 12. Ed. A. L. Muggleton-Harris, Academic Press, New York, 1981, pp. 113–176.

Wedner, H. J. and Parker, C. W. *Biochem. J.* **162**:483–491 (1977).

Wehland, J.; Willingham, M. C.; Dickson, R.; and Pastan, I. *Cell* **25**:105–119 (1981).

Weigele, J. B. and Barchi, R. L. *Proc. Natl. Acad. Sci.,* U.S. **79**:3651–3655 (1982).

Weinberg, C. B. and Hall, Z. W. *Proc. Natl. Acad. Sci.,* U.S. **76**:504–508 (1979).

Weinberg, C. B.; Reiness, C. G.; and Hall, Z. W. *J. Cell Biol.* **88**:215–218 (1981).

Weir, M. P. and Lo, C. W. *Proc. Natl. Acad. Sci.,* U.S. **79**:3232–3235 (1982).

Weiss, M. J. and Luria, S. E. *Proc. Natl. Acad. Sci.,* U.S. **75**:2483–2487 (1978).

White, J. and Helenius, A. *Proc. Natl. Acad. Sci.,* U.S. **77**:3273–3277 (1980).

Wickner, W. *Science* **210**:861–868 (Nov. 1980).

Widnell, C. C.; Schneider, Y. J.; Pierre, B.; Baudhuin, P.; and Trouet, A. *Cell* **28**:61–70 (1982).

Wilkerson, L.; Perkins, R., Jr.; Roelofs, R.; Swift, L.; Dalton, L.; and Park, J. *Proc. Natl. Acad. Sci.,* U.S. **75**:838–841 (1978).

Willard, A. L. *J. Physiol.* **301**:115–128 (1980).

Williams, D. G.; Jenkins, R. E.; and Tanner, M. J. A. *J. Biochem.* **181**:477–493 (1979).

Willingham, M. C. and Pastan, I. *J. Cell Biol.* **67**:146–159 (1975).

Willingham, M. C. and Pastan, I. *Cell* **21**:67–77 (1980).

Wilson, R. J. *Nature,* London **295**:368–369 (1982).

Wise, D. S.; Karlin, A.; and Schoenborn, B. P. *Biophys. J.* **28**:473–496 (1979).

Wise, D. S.; Schoenborn, B. P.; and Karlin, A. *J. Biol. Chem.* **256**:4124–4126 (1981).

Witkowski, J. A. and Jones, G. E. *Trends Biochem. Sci.* **6**:IX–XII (1981).

Wolosewick, J. J. and Porter, K. R. *J. Cell Biol.* **82**:114–139 (1979).

Wood, J. G.; Wallace, R. W.; Whitaker, J. N.; and Cheung, W. Y. *J. Cell Biol.* **84**:66–76 (1980).

Wood, R. L. *J. Ultrastructure Res.* **58**:299–315 (1977).

Woodward, M. P. and Roth, T. F. *Proc. Natl. Acad. Sci.*, U.S. **75**:4394–4398 (1978).

Wu, C. F. and Ganetzky, B. *Nature,* London **286**:814–817 (1980).

Wurster, B.; Pan, P.; Tyan, G. G.; and Bonner, J. T. *Proc. Natl. Acad. Sci.*, U.S. **73**:795–799 (1976).

Yamada, K. M. and Olden, K. *Nature,* London **275**:179–184 (1978).

Yu, J. and Branton, D. *Proc. Natl. Acad. Sci.*, U.S. **73**:3891–3895 (1976).

Yunghans, W. N. and Morré, D. J. *Cytobiologie* **17**:212–231 (1978).

Zaccai, G. and Gilmore, D. J. *J. Mol. Biol.* **132**:181–191 (1979).

Zingsheim, H. P.; Neugebauer, D. C.; Barrantes, F. J.; and Frank, A. *Proc. Natl. Acad. Sci.*, U.S. **77**:952–956 (1980).

Index